NJU SA 2018-20**19**

南京大学建筑与城市规划学院　　建筑学教学年鉴

THE YEAR BOOK OF ARCHITECTURE TEACHING PROGRAM

SCHOOL OF ARCHITECTURE AND URBAN PLANNING NANJING UNIVERSITY

王丹丹 编　　EDITOR：WANG DANDAN

东南大学出版社·南京　SOUTHEAST UNIVERSITY PRESS，NANJING

U0179553

建筑设计及其理论
Architectural Design and Theory

张 雷 教 授	Professor ZHANG Lei
冯金龙 教 授	Professor FENG Jinlong
吉国华 教 授	Professor JI Guohua
周 凌 教 授	Professor ZHOU Ling
傅 筱 教 授	Professor FU Xiao
王 铠 副教授	Associate Professor WANG Kai
钟华颖 副教授	Associate Professor ZHONG Huaying
黄华青 副教授	Associate Professor HUANG Huaqing

城市设计及其理论
Urban Design and Theory

丁沃沃 教 授	Professor DING Wowo
鲁安东 教 授	Professor LU Andong
华晓宁 副教授	Associate Professor HUA Xiaoning
胡友培 副教授	Associate Professor HU Youpei
窦平平 副教授	Associate Professor DOU Pingping
刘 铨 副教授	Associate Professor LIU Quan
唐 莲 副教授	Associate Professor TANG Lian
尤 伟 副教授	Associate Professor YOU Wei
尹 航 讲 师	Lecturer YIN Hang

建筑历史与理论及历史建筑保护
Architectural History and Theory, Protection of Historic Building

赵 辰 教 授	Professor ZHAO Chen
王骏阳 教 授	Professor WANG Junyang
胡 恒 教 授	Professor HU Heng
冷 天 副教授	Associate Professor LENG Tian
史文娟 副教授	Associate Professor Shi Wenjuan

建筑技术科学
Building Technology Science

鲍家声 教 授	Professor BAO Jiasheng
吴 蔚 副教授	Associate Professor WU Wei
郜 志 副教授	Associate Professor GAO Zhi
童滋雨 副教授	Associate Professor TONG Ziyu
梁卫辉 副教授	Associate Professor LIANG Weihui
孟宪川 讲 师	Lecturer MENG Xianchuan
施珊珊 讲 师	Lecturer SHI Shanshan

南 京 大 学 建 筑 与 城 市 规 划 学 院 建 筑 系
Department of Architecture
School of Architecture and Urban Planning
Nanjing University
arch@nju.edu.cn   http://arch.nju.edu.cn

教学纲要
EDUCATIONAL PROGRAM

教学阶段 Phases of Education	本科生培养（学士学位）Undergraduate Program (Bachelor Degree)			
	一年级 1st Year	二年级 2nd Year	三年级 3rd Year	四年级 4th Year

教学类型 Types of Education	通识教育 General Education			专业教育 Professional

课程类型 Types of Courses	通识类课程 General Courses	学科类课程 Disciplinary Courses		专业类课程 Professional Cours

主干课程 Design Courses	设计基础 Basic Design	建筑设计基础 Basic of Architectural Design	建筑设计 Architectural Design	

理论课程 Theoretical Courses	基础理论 Basic Theory of Architecture	专业理论 Architectural Theory		

技术课程 Technological Courses				

实践课程 Practical Courses	环境认知 Environmental Cognition	古建筑测绘 Ancient Building Survey and Drawing	工地实习 Practice of Construction Plant	

研究生培养（硕士学位）Graduate Program (Master Degree)			研究生培养（博士学位）Ph. D. Program(Doctoral Degree)
一年级 1st Year	二年级 2nd Year	三年级 3rd Year	

学术研究训练 Academic Research Training

学术研究
Academic Research

建筑设计研究 Research of Architectural Design	毕业设计 Thesis Project	学位论文 Dissertation	学位论文 Dissertation
专业核心理论 Core Theory of Architecture	专业扩展理论 Architectural Theory Extended	专业提升理论 Architectural Theory Upgraded	跨学科理论 Interdisciplinary Theory

构造实验室 Tectonic Lab

建筑物理实验室 Building Physics Lab

数字建筑实验室 CAAD Lab

生产实习 Practice of Profession	生产实习 Practice of Profession

课程安排
CURRICULUM OUTLINE

	本科一年级 Undergraduate Program 1st Year	本科二年级 Undergraduate Program 2nd Year	本科三年级 Undergraduate Program 3rd Year
建筑设计 Architectural Design	设计基础 Basic Design	建筑设计基础 Basic of Architectural Design 建筑设计（一） Architectural Design 1 建筑设计（二） Architectural Design 2	建筑设计（三） Architectural Design 3 建筑设计（四） Architectural Design 4 建筑设计（五） Architectural Design 5 建筑设计（六） Architectural Design 6
建筑理论 Architectural Theory		建筑导论 Introductory Guide to Architecture 中国传统建筑文化 Traditional Chinese Architecture	建筑设计基础原理 Basic Theory of Architectural Design 居住建筑设计与居住区规划原理 Theory of Housing Design and Residential Planning 城市规划原理 Theory of Urban Planning
建筑技术 Architectural Technology	理论、材料与结构力学 Theoretical, Material & Structural Statics Visual Basic程序设计 Visual Basic Programming	CAAD理论与实践（一） Theory and Practice of CAAD 1	建筑技术（一）：结构、构造与施工 Architectural Technology 1: Structure, Construction & Executio 建筑技术（二）：建筑物理 Architectural Technology 2: Building Physics 建筑技术（三）：建筑设备 Architectural Technology 3: Building Equipment
历史理论 History Theory		外国建筑史（古代） History of World Architecture (Ancient) 中国建筑史（古代） History of Chinese Architecture (Ancient)	外国建筑史（当代） History of World Architecture (Modern) 中国建筑史（近现代） History of Chinese Architecture (Modern)
实践课程 Practical Courses		古建筑测绘 Ancient Building Survey and Drawing	工地实习 Practice of Construction Plant
通识类课程 General Courses	数学 Mathematics 语文 Chinese 思想政治 Ideology and Politics 科学与艺术 Science and Art	社会学概论 Introduction of Sociology	
选修课程 Elective Courses		城市道路与交通规划 Planning of Urban Road and Traffic 环境科学概论 Introduction of Environmental Science 人文科学研究方法 Research Method of the Social Science 管理信息系统 Management Operating System 城市社会学 Urban Sociology	人文地理学 Human Geography 中国城市发展建设史 History of Chinese Urban Development 欧洲近现代文明史 Modern History of European Civilization 中国哲学史 History of Chinese Philosophy 宏观经济学 Macro Economics

本科四年级 Undergraduate Program 4th Year	研究生一年级 Graduate Program 1st Year	研究生二、三年级 Graduate Program 2nd & 3rd Year
建筑设计（七） Architectural Design 7 建筑设计（八） Architectural Design 8 本科毕业设计 Graduation Project	建筑设计研究（一） Design Studio 1 建筑设计研究（二） Design Studio 2 数字建筑设计 Digital Architecture Design 联合教学设计工作坊 International Design Workshop	专业硕士毕业设计 Thesis Project
城市设计及其理论 Urban Design and Theory 都市社会学 Urban Sociology	中国木建构文化研究 Studies in Chinese Wooden Tectonic Culture 城市形态与建筑设计方法论 Urban Form and Architectural Design Morphology 现代建筑设计基础理论 Preliminaries in Modern Architectural Design 现代建筑设计方法论 Methodology of Modern Architectural Design 研究方法与写作规范 Research Method and Thesis Writing	
建筑师业务基础知识 Introduction of Architects' Profession CAAD理论与实践（二） Theory and Practice of CAAD 2	材料与建造 Materials and Construction 计算机辅助建筑设计技术 Technology of CAAD GIS基础与运用 Concepts and Application of GIS	建筑环境学与设计 Architectural Enviromental Science and Design 建筑技术中的人文主义 Technology Humanism in Architecture
	建筑理论研究 Studies of Architectural Theory	
生产实习（一） Practice of Profession 1	生产实习（二） Practice of Profession 2	建筑设计与实践 Architectural Design and Practice
景观规划设计及其理论 Landscape Planning Design and Theory 地理信息系统概论 Introduction of GIS 欧洲哲学史 History of European Philosophy 宏观经济学 Macro Economics 建筑节能与绿色建筑设计 Building Energy Efficiency and Green Building Design	景观都市主义理论与方法 Theory and Method of Landscape Urbanism 建筑史研究 Studies in Architectural History 建筑体系整合 Advanced Building System Integration 算法设计 Algorithmic Design 建筑环境学 Architectural Environmental Science 传统热力学与计算流体力学基础 Fundametals of Heat Transfer and Computational Fluid Dynamics	建设工程项目管理 Management of Construction Project

1—127　年度改进课程　WHAT'S NEW

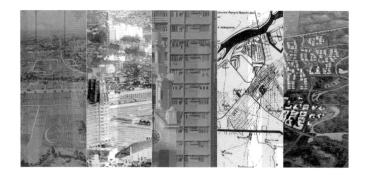

年度改进课程
WHAT'S NEW

设计基础
BASIC DESIGN

鲁安东 唐莲 尹航 黄华青

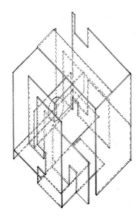

1.课程背景

在南京大学的通识教育体系下，本科生一年级第二学期的"设计基础"作为设计相关学科的启蒙课程，既是本科大类通识课的组成部分，也是建筑与城市规划学院工科教育"三三制"的重要准备阶段。课程除了面向准备进入建筑学、城乡规划专业的学生外，还面向部分通识性质的其他专业学生。因此，课程设置的最大挑战便是，如何在大类通识、工科通识与建筑学专业教育的诉求和预期之间取得平衡，在趣味性和专业性之间取得平衡。本课程以艺术、人文、科学、工程的综合素质教育为目标，结合建筑学专业的基础能力培养，试图探索一种宽口径、多样化的"新工科"设计课程体系。

课程采取了三个模块循序渐进、多个小组自主命题的新模式。三个模块分别为：A-感知（转化能力训练）、B-分析（制图能力训练）、C-创造（动手能力训练），旨在引导学生形成建成环境的初步认知，掌握建筑设计的初步技能。每个模块分为四个横向小组，由不同兴趣方向的教师在体系框架下充分发挥命题的自主性和创新性——包括人/环境、诗意、尺度等若干不同主题而又相互关联的子课程，由此编织成一张立体的"知识网格"。每个模块的教学时间为五周，各模块最后一周为交流评图。

2.典型案例

以下以其中一条课程线索的三个模块为例。

理解栖居的诗意，营造诗意的栖居，是建筑学的基本议题之一。其背后的核心，即是对空间/形式与人的关联性的探讨：如何营造一个能够容纳人、容纳人的生活的空间，以实现栖居诗意的体化。它作为一个基本的建筑学话题，可承载从业余至专业领域不同程度的感知、分析和创造活动；该议题同时是个交叉学科的话题，融汇了哲学、人文、艺术、科学、技术等领域的思考——这对于本课程同时作为通识教育基础课以及建筑学专业准入课的定位而言，是一个恰当的切入点。

课程提出诗意栖居空间的三个基本属性：叙事性（narrative）、日常性（everyday）、原形性（prototypical）。由此构筑一条渐次推进的学习路径，形成三个设计模块："书写空间""日常空间"与"聚落空间"。三个模块以"感知—分析—创造"的课程总纲为线索，以空间/形式这一对核心关系为根基，着重训练学生在空间美学、空间结构以及形式创造层面的基本素质。

2.1 书写空间

"书写空间"模块从大一学生熟悉的文字诗意开始，搭建基于文本的美学鉴赏和空间的美学感知之间的桥梁。明清江南私家园林作为一种融于生活的"整体艺术"，提供了感知诗意栖居空间的完美对象；而明清造园典籍以及文人创作的园记、园林图等，则作为引导学生进入园林美学体系的文本。

本模块包括两个部分。首先是"从空间到文本"：学生选取某个现实园林，对其中一个场景进行深入观察，撰写一段与空间感受相关的文字，提炼空间美学要素；接着借助《长物志》等典籍，对园林场景中的一类特定生活美学元素（如室庐、花木、水石、书画等）进行鉴赏与分析。

然后是"从文本到空间"，以《止园记》为切入点，提取文字中的空间元素和空间关系，进而采用现代建筑学语汇，抽象呈现园林中的空间感受。最后的模型表达，学生以规定数量的白色板片，在18 cm×18 cm×18 cm立方体的盒子内，重现《止园记》中书写的诗意生活场景，如《止园记》中提到的水周堂闻桂、鸿磬轩品茗、梨云楼赏花等。学生在文本与空间的反复穿梭过程中，逐步摸索栖居诗意在建成环境中的叙事表达。

2.2 日常空间

"日常空间"模块旨在引导学生从知识性的感知走向实践性的观察，挖掘日常生活的诗意与创造，并初步学习复杂建成环境的分析方法和图示化表达。该模块以"空间—社会"研究方法为基础，以南京某城中村为地段，要求学生深入城中村日常空间的不同维度，探索日常生活实践与空间的互动关系。

本模块包括三个作业。首先是"日常空间观察"，每组学生选取城中村的一个公共空间节点，进行长时段观察，借助照片和图纸记录空间中的典型行为及其与空间的关系；其次是"日常空间阅读"，将空间节点投射在聚落空间中，以图纸的方式思考和表达"城—村"关系；最后一步由"外"到"内"，选择城中村的一座多院落民居，通过观察、访谈和记录，试图理解居住人群的社会结构（年龄、职业、亲属关系等），完成一个能够展现民居中日常生活空间及社会关系的实体模型。

模型成果呈现了日常生活的多样性，学生选择了截然不同的观察视角，如杂院中充满公共活动的狭长巷道、杂院中房东与房客的空间区隔、三口之家的精心生活等等。学生观察生活的眼光一定程度上得到了磨炼。

2.3 聚落空间

"聚落空间"模块则在类型学及形态学框架内，引导学生综合地想象诗意栖居的

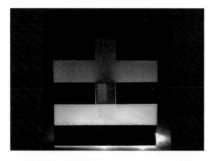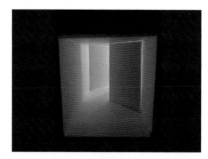
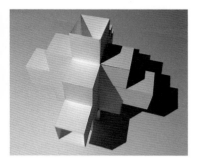

城市与建筑空间，进行基于空间原形的设计探索。研究案例从日常空间拓展至更广泛而复杂的"世界聚落"之中，旨在提供更具典型性、理想化的类型学和形态学样本。

本模块包括三个作业。首先是"聚落空间图式"，学生从教师提供的世界聚落样本库中选择特定的聚落类型，绘制图底关系图，每组聚落皆具有明显的空间元素特征，如广场、墙、路、桥、阶、塔等。接下来的"聚落空间元素"作业，学生需针对某个聚落空间元素与日常生活的关系，完成一个聚落空间序列的叙事规划。最后，基于对聚落空间原形的分析，学生在指定范围内（100 cm×50 cm×50 cm）搭建一座满足未来生活想象的"看不见的聚落"。这一原形城市凝聚了学生对于诗意栖居的空间理解，并呈现出垂直城市的形态想象，也赋予了学生较大的自由度。

3.成果与反思

通过以上三个模块的练习，学生逐步理解并学会建构人与环境的关系，从最熟悉的"人"的研究，走向一种营造诗意栖居的空间形式。

课程设置整体上兼具专业性和综合性，不仅涉及广泛的建成环境学科议题，也对建筑学的基本技能进行了初步训练。为了应对通识教育与专业教育的平衡问题，课程采取了从要素到整体的教学策略，无论是书写空间的美学元素、日常空间的生活要素还是聚落空间的空间原形要素，教师帮助和引导学生对复杂建成环境进行了元素分解和凝炼工作，降低了学生初步尝试建筑设计的难度，让学生获得更多成就感。尽管每个模块的教学进程都比较紧张，但对最终的成果要求并未降低，学生的潜能得到了充分挖掘。当然，课程难度设置上多少超出了大部分学生的能力范畴，学生之间差距较为明显。

以诗意栖居为主题，不仅兼顾了通识教育的综合诉求，也完成了与建筑学教育的衔接工作。建筑学训练为工科大类学生提供了基础美学素养和技术常识；同时，通识教育的视野亦反过来拓宽了建筑学的边界，使建筑学跃出按部就班的传统工科教育路径，以探索一种宽口径、多样化的新时代人才培养新模式。这一探索之路刚刚开始。

1.Course background

Under the general education system of Nanjing University, as an enlightening course of design-related disciplines, "design basis" for undergraduates in the second semester of freshman year is not only an undergraduated integral part of general education, but also an important preparatory stage for the "three-three-system" of engineering education in the School of Architecture and Urban Planning. In addition to students preparing to enter architecture, urban planning majors, there are also some general-purpose students in other majors. Therefore, the biggest challenge of course design is how to strike a balance between the demands and expectations of general education, engineering general education and architectural professional education, and between interest and professionalism. This course aims at the comprehensive quality education of art, humanities, science and engineering, and tries to explore a wide-caliber and diversified "new subject" design curriculum system, combining with the basic ability training of architecture specialty.

The course adopts a new model of three modules step by step and multiple group independent propositions. The three modules are: A-perception (transformation ability training), B-analysis (cartography ability training) and C-creation (hands-on ability training). The purpose is to guide students to form a preliminary understanding of the built environment and master the preliminary skills of architectural design. Each module is divided into four horizontal groups. Teachers with different interest directions give full play to the autonomy and innovation of propositions under the framework of the system, including several sub-courses on different subjects such as human/environment, poetry/scale and so on, which weave into a three-dimensional "knowledge grid". The teaching time of each module is five weeks, and the last week of each module is to communicate and score charting.

2.Typical cases

Here take three modules of one of the course clues as an example.

Understanding the poetry of residence and creating a poetic residence is one of the basic topics of architecture. The core behind it is to explore the relationship between space/form and people: how to create a space that can accommodate people and their lives so as to realize the embodiment of poetic residence. As a basic architectural topic, it can carry different degrees of perception, analysis and creative activities from amateur to professional fields; at the same time, this topic is an interdisciplinary topic, which integrates many fields of thinking, such as philosophy, humanities, art, science, technology and so on—It is an appropriate entry point for the orientation of this course as a basic course of general education as well as an admission course of architecture specialty.

The course puts forward three basic attributes of poetic residence space: narrative, everyday and prototypical. Thus, a progressive learning path is constructed, and three design modules are formed, namely "writing space""daily space" and "settlement space". The three modules regard the general curriculum of "perception-analysis-creation" as clue and are based on the core relationship of space/form, focus on training students' basic qualities at the level of space aesthetics, space structure and form creation.

2.1 Writing space

The module of "writing space" begins with the familiar characters poetry to freshmen, and builds a connection between aesthetic appreciation based on text and aesthetic perception of space. The private gardens in the south region of the Yangtze River in the Ming and Qing Dynasties, as a "whole art" blending with life, provide the perfect object for perceiving the poetic living space; while the garden classics in the Ming and Qing Dynasties, as well as the garden records and garden paintings created by literati, are served as texts to guide students into the aesthetic system of gardens.

This module includes two parts. Firstly, "from space to text": Students select a real garden, observe one scene in depth, write a paragraph of text related to space experience, and refine the elements of space aesthetics; then with the help of *Changwuzhi* and other classics, appreciate and analyse a kind of specific aesthetic elements of life in garden scenes, such as houses, flowers and trees, water and stone, calligraphy and painting, etc.

Then it is "from text to space", taking the garden diary as the breakthrough point, extracting the space elements and spatial relations in the text, and then using modern architectural vocabulary to abstractly present the space feeling in the garden. In the final model, the students reproduce the poetic scenes of life written in the garden records in the box of 18 cm×18 cm×18 cm cube with a specified number of white plates, such as smell osmanthus fragrance in Shuizhou Hall, taste tea in Hongqing Xuan and appreciate flowers in Liyun Tower mentioned in *Zhiyuanji*. During the repeated shuttle between text and space, students gradually explore the narrative expression of dwelling poetry in the built environment.

2.2 Daily space

"Daily space" module aims to guide students from knowledge perception to practical observation, excavate the poetry and creation of daily life, and preliminarily learn the analytical methods and graphical expression of complex built environment. This module is based on the "space-society" research method and takes an urban village in Nanjing as a section. It requires students to go deep into the different dimensions of daily space in the urban village and explore the interaction between daily life practice and space.

This module includes three works. The first is "daily space observation". Each group of students selects a public space node of the urban village to observe for a long period of time, and records typical behavior in the space and its relationship with space by means of photos and drawings. The second, is "daily space reading", projecting space nodes into settlement space, thinking and expressing "city-village"

relationship in the way of drawings. The last step is from "outside" to "inside", choosing a multi-courtyard residence in an urban village. Through observation, interviews and records, try to understand the social structure of the residents (age, jobs, relatives, etc.) and complete an entity model that can show the daily living space and social relations in the dwelling.

The results of the model show the diversity of daily life. Students choose different perspectives of observation, such as narrow lanes full of public activities in miscellaneous courtyards, space separation between landlords and tenants in miscellaneous courtyards, meticulous life of three families, etc. To a certain extent, the students' vision of observing life has been tempered.

2.3 Settlement space

In the framework of typology and morphology, the "settlement space" module guides students to imagine the urban and architectural space of poetic residence, and to explore the design based on spatial prototype. The case study extends from everyday space to a wider and more complex world settlement, aiming at providing more typical and idealized typological and morphological samples.

This module includes three works. The first is the "settlement space schema". Students select a specific settlement type from the sample bank of the world settlement provided by the teachers and draw a map of the relationship between the base. Each group of settlements has obvious characteristics of spatial elements, such as square, wall, road, bridge, step, tower, etc.. Next, in the assignment of "settlement space elements", students need to complete a narrative planning of the sequence of settlement space in view of the relationship between a settlement space element and daily life. Finally, based on the analysis of the prototype of settlement space, students build an "invisible settlement" within a specified range (100 cm × 50 cm × 50 cm) to satisfy the imagination of future life. This prototype city embodies the students' spatial understanding of poetic residence, presents the image of vertical city, and also gives students greater freedom.

3.Achievements and reflections

Through the practice of the above three modules, students gradually understand and learn to construct the relationship between human and environment, from the most familiar "human" research to a space form of creating a poetic residence.

As a whole, the curriculum of course design is both professional and comprehensive, involves not only a wide range of issues of building environment discipline, but also a preliminary training of basic skills of architecture. In order to cope with the balance between general education and professional education, the curriculum adopts a teaching strategy from elements to the whole, whether it is the aesthetic elements of writing space, the living elements of daily space or the prototype elements of settlement space, teachers help and guide students to decompose and condense elements of complex built environment, which reduces the difficulty of students' preliminary attempt of architectural design, and enables students to get more sense of achievement. Although the teaching process of each module is rather tense, the requirements for the final results have not been reduced, and the potential of students has been fully tapped. Of course, the difficulty of course setting is more or less beyond the scope of most students' abilities, and the gap between students is more obvious.

The theme of poetic residence not only takes into account the comprehensive demands of general education, but also completes the work of linking up with architectural education. Architecture training provides basic aesthetic literacy and technical common sense for engineering students; at the same time, the vision of general education has also broadened the boundaries of architecture, enabling architecture to leap out of the traditional engineering education path, in order to explore a wide-caliber, diversified new era of talent training mode. This road of exploration has just begun.

感受认知	分析转化	创造设计
5周（个人作业）	5周（个人作业）	5周（小组作业）
每组20人左右	每组20人左右	每组20人左右

A1 鲁安东
认识"关系"

B1 鲁安东
园林剧场

C1 鲁安东
电影建筑

A2 唐莲
度量空间

B2 唐莲
游历空间

C2 唐莲
包裹空间

A3 黄华青
书写空间

B3 黄华青
日常空间认知

C3 黄华青
聚落空间重构

A4 尹航
城市空间认知

B4 尹航
建筑空间认知

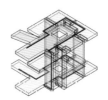

C4 尹航
城市街道界面分析

C1: 电影建筑
C1:FILM ARCHITECTURE
鲁安东

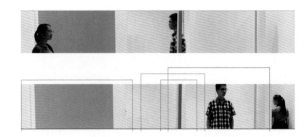

1.教学目标

该作业由两段影像成果构成,核心是认识和表现空间与人之间的密切关系,即空间"激发"了人的特定活动,而人的活动"演出"了空间的潜力。

第一段影像以"寻找"为主题,从二位演员之间不断变化的空间关系入手,为演员的空间调度搭建了一个临时的空间场景,并用一个固定镜头加以记录。

第二段影像以"蹑踪"为主题,研究校园内的一段空间并分析其具有的叙事潜力,进而为这一空间片段量身定制一段空间剧情,随后通过二位演员的表演呈现空间内在的叙事结构特征,并用一个长镜头加以记录。

作业由四名学生合作完成,分工包括导演、演员和摄像,使学生通过分析、表演和拍摄等多重视角,全面理解空间中人的活动。

2.教学内容

本模块通过拍摄电影的形式帮助学生循序渐进地建立一种观察和认识空间的方式,培养学生的空间感知和空间想象的能力。本作业包括三个练习:

(1) 空间中的观察:二人一组,从多个角度再现空间中的一段运动。

(2) 空间中的运动:四人一组,单一(运动)镜头表达空间中的人物关系。

(3) 空间中的叙事:四人一组,单一(运动)镜头表达空间中的人物关系。

3.教学进程

第1周

(1) 课上练习A(二人组):空间中的观察。

(2) 学生二人一组,设计一段空间运动,选四张照片记录这段运动中的关键变化。

第2周

(1) 课上汇报拍摄方案(课上点评并二选一进行拍摄)。

(2) 课上练习B(四人组):空间中的运动。

第3周

(1) 课上汇报练习B拍摄成果。

(2) 课上PPT汇报作业C场地和拍摄方案,点评讨论。

第4周

评图。

1.Training objective

This assignment consists of two videos, with the core of recognizing and expressing the close relationship between space and people, that is, space "stimulates" people's specific activities, and people's activities "perform" the potential of space.

The first video is themed in "looking for", and starts with the ever-changing spatial relationship between the two actors. A temporary space scene is built for the actors' spatial dispatching and a fixed shot is used to record.

The second video is themed in "following", studies a space in the campus and analyzes its narrative potential, so as to customize a space plot for this space segment, then presents the internal narrative structure characteristics of the space through the performance of two actors, and records it with a full-length shot.

The assignment is completed by the cooperation of four students, including director, actor and camera shooting, so that students can comprehensively understand the activities of people in the space through multiple perspectives such as analysis, performance and shooting.

2.Teaching content

Introduction: this module helps students gradually establish a way to observe and understand space by shooting a film, and cultivates students' ability of spatial perception and imagination.

3.Teaching process

The first week

(1) Class exercise A (2 persons in a group): observation in space.

(2) 2 students in a group, design a spatial movement, choose 4 photos to record the key changes in the movement (1.5 h).

The second week

(1) Report shooting plan in the class (make comments in the class and choose 1 for shooting) (1 h).

(2) Class exercise B (4 persons in a group): movement in space.

The third week

(1) Report exercise B shooting results in the class (1 h).

(2) Report site and shooting plan of assignment C through PPT in the class, make comments and discuss.

The fourth week

Drawing evaluation.

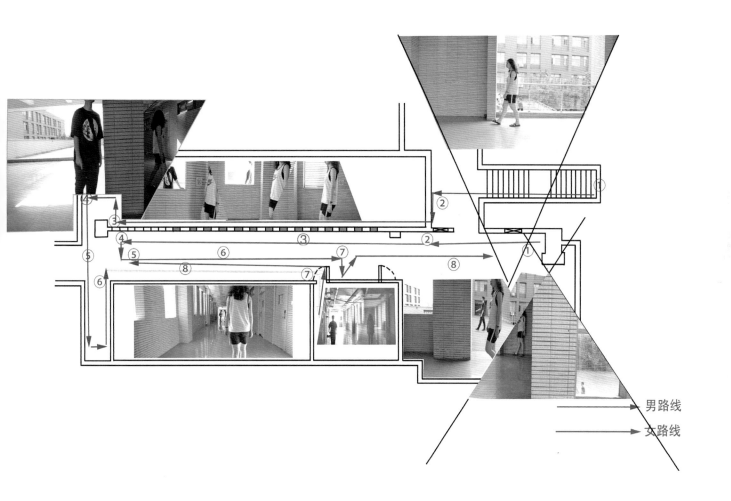

① ② ③ ④ ⑤ ⑥ ⑦ ⑧

男路线
女路线

开端
Opening

发展
Development

高潮
Climax

结局
Ending

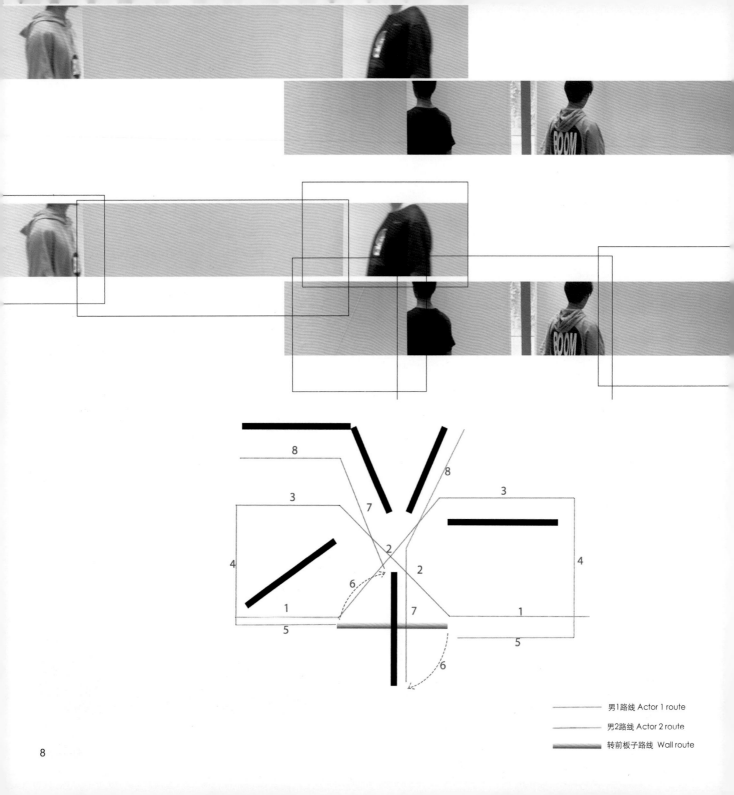

男1路线 Actor 1 route

男2路线 Actor 2 route

转前板子路线 Wall route

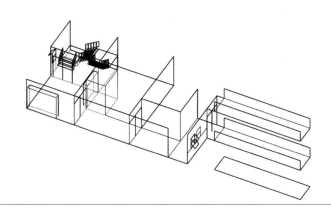

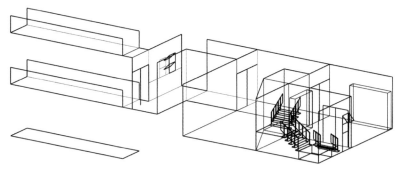

C2: 包裹空间
C2: WRAPPING SPACE
唐莲

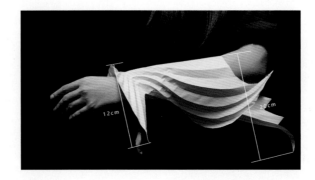

1.教学目标

理解形式塑造机制，理解形式与材料、构件、工艺的关系。

2.教学内容

"包裹空间"的教学历时五周，要求用折纸对身体部位进行包裹，完成设计与制作。课程可以理解为基于身体（场地）的形式操作，教学的主要内容是形式设计的逻辑与方法，其中折纸作为实现形式的技术与媒介。为此，在整个教学过程中设置了三个阶段的练习，并开展相应的讲座来指导与配合练习。这三个阶段分别为：折纸单元基础练习（1周）、折纸单元变形与组合研究（1周）以及折纸包裹空间的设计（3周）。

3.教学进程

第十二周

1～2人一组，折纸单元的变形与组合原理研究。

（1）课程介绍

（2）课上练习：1～2人一组，选择折纸单元，进行折纸单元的尝试。

（3）讲课："折纸单元的变形与组合"（唐莲）。

第十三周

二人一组，选择包裹部位，提出设计概念，完成初步设计。

（1）课后作业讲评：扫描合成图（电子版）。

（2）讲课："身体尺度与空间包裹"（唐莲）。

（3）课上练习：选择身体部位，进行几何分析。

第十四周

（1）课后作业讲评。

（2）讲课："图示表达"（唐莲）。

第十五周

完成制作，完成图纸：手绘分析设计作品的几何构成与拍摄折叠过程成品。

第十六周

答辩。

1.Training objective

Understand form shaping mechanism and the relationship between form and materials, components and processes.

2.Teaching content

The teaching of "wrapping space" lasts for five weeks. Students are required to use paper folding to wrap body parts to complete design and production. The course can be understood as the form operation based on body (site). The main content the teaching is the logic and method of form design, among which paper folding the technology and medium to realize the form. Therefore, three stages of exercise are set in the whole teaching process, and relevant lectures are held to guide a coordinate the exercises. The three stages are: basic exercise of paper folding u (1 week), study on deformation and combination of paper folding unit (1 week), a design of paper folding wrapping space (3 weeks).

3.Teaching process

The 12th week

1~2 persons in a group, study on deformation and combination principles of pap folding unit.

(1) Introduction.

(2) Class exercise: 1~2 persons in a group, choose paper folding unit, try pap folding unit.

(3) Lecture: Deformation and Combination of Paper Folding Unit (Tang Lian).

The 13th week

2 persons in a group, choose wrapping parts, propose design concept, comple preliminary design.

(1) Comments on homework: scan composite graph (electronic version).

(2) Lecture: Body Size and Space Wrapping (Tang Lian).

(3) Class exercise: choose body parts for geometrical analysis.

The 14th week

(1) Comments on homework.

(2) Lecture: Graphical Expression (Tang Lian).

The 15th week

Complete production, complete drawing: hand draw and analyze geometric composition of design works and take photos of finished folding process product.

The 16th week

Defense.

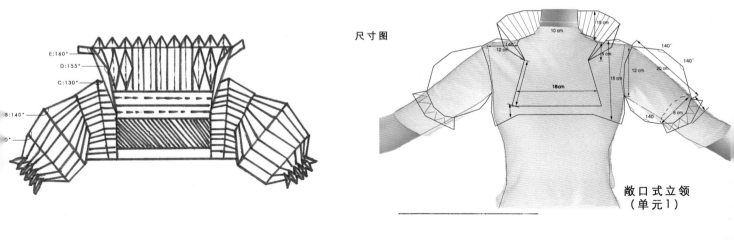

尺寸图

10 cm

15 cm

12 cm

140°

5 cm

140°

140°

18cm

15 cm

12 cm

20 cm

140

6 cm

E:160°

D:155°

C:130°

B:140°

D:

敞口式立领
（单元1）

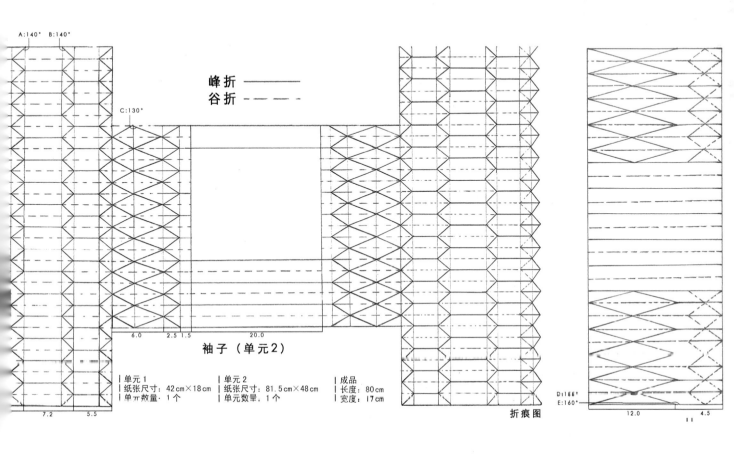

A:140°   B:140°

峰折 ————
谷折 — — —

C:130°

6.0   2.5 1.5   20.0

袖子（单元2）

7.2   5.5

D:155°
E:160°

12.0   4.5

折痕图

单元 1	单元 2	成品
纸张尺寸：42 cm×18 cm	纸张尺寸：81.5 cm×48 cm	长度：80 cm
单元数量：1 个	单元数量：1 个	宽度：17 cm

11

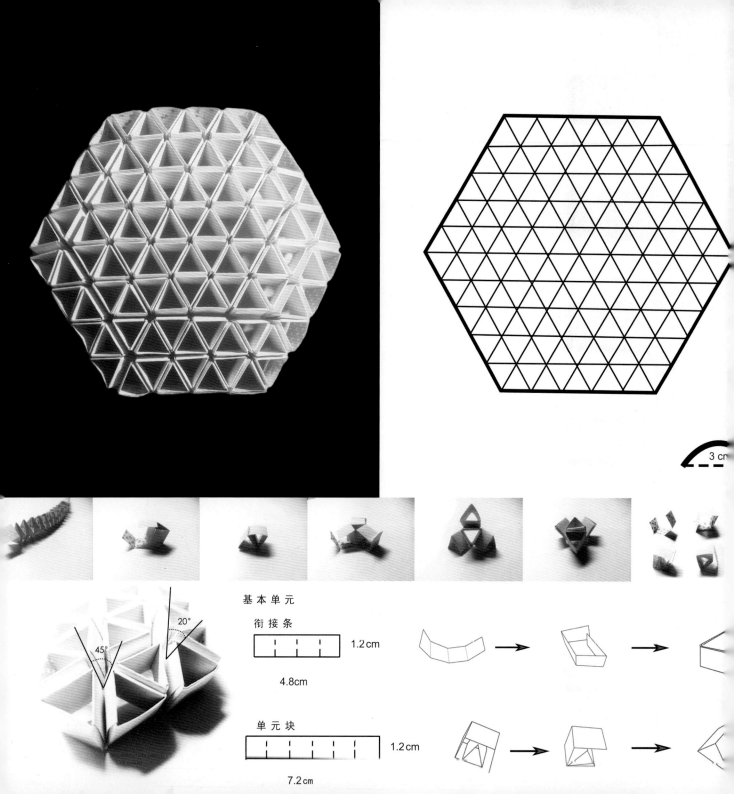

3 cm

基 本 单 元

衔 接 条

1.2 cm

4.8 cm

单 元 块

1.2 cm

7.2 cm

20°

45°

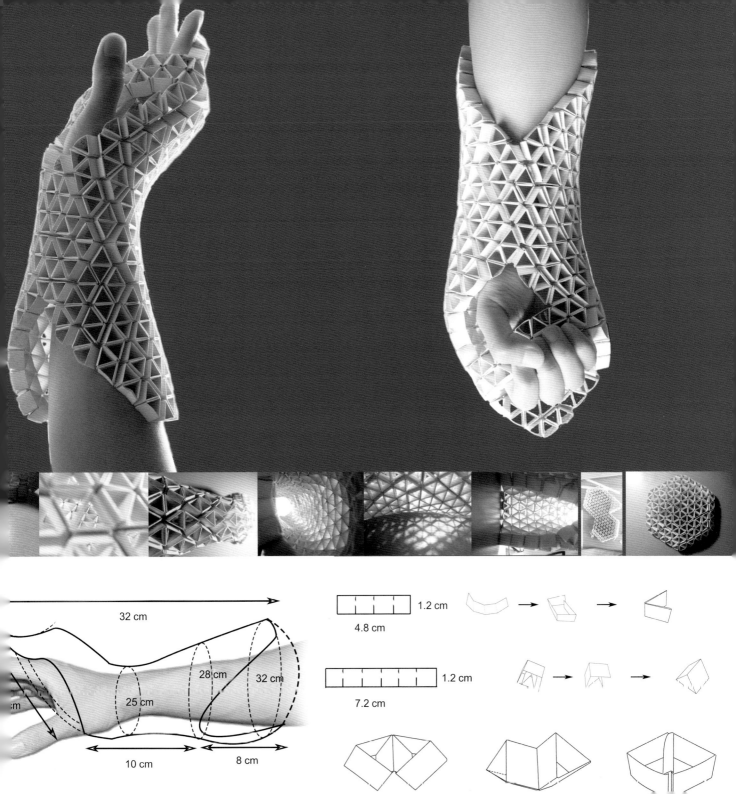

32 cm

28 cm

32 cm

25 cm

10 cm

8 cm

1.2 cm

4.8 cm

1.2 cm

7.2 cm

C3: 聚落空间重构
C3: RECONSTRUSTION OF SETTLEMENT SPACE
黄华青

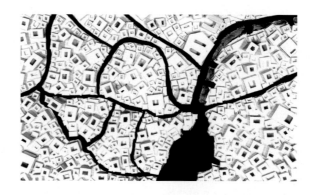

1.教学目标

培养学生对聚落形态及空间类型的分析和重构能力，运用空间原形进行初步的城市和建筑的空间想象和设计。

2.教学内容

"聚落重构"模块选取国内外典型聚落为研究对象，指导学生借助地图软件及文献资料，从形态学和类型学视角提取聚落空间的组成要素，并基于聚落空间原形设计一座"理想之城"。教学过程历时四周（含评图一周），包括三个练习：

（1）一人一组，完成一个典型聚落形态的图底关系分析。

（2）二人一组，针对一个典型聚落空间元素，完成一个聚落空间序列的叙事表达模型。

（3）二人一组，基于聚落空间原形，在给定框架内设计一座"理想之城"。

3.教学进程

第一周

聚落空间图示

（1）讲课："世界聚落的教示"。

（2）选取一个特定类型的典型聚落，借助地图软件和相关文献，绘制聚落的图底关系分析图。

第二周

聚落空间元素

（1）在聚落分析图基础上，针对一个聚落空间元素与日常生活的关系，完成一个聚落空间序列的叙事策划。

（2）制作能够表达该聚落空间序列的建筑模型。

第三周

聚落空间重构（模型）

（1）基于特定的聚落空间元素和原形，在给定范围内（如100 cm×50 cm×50 cm），设计一座满足未来生活想象的"理想之城"，一座"看不见的聚落"。

（2）深化制作模型，汇总最终成果。

第四周

答辩。

1.Traning objective

Cultivate students' ability to analyze and reconstruct settlement forms and spatia[l] types, and use spatial prototypes to have preliminary space imagination and design c[f] city and architecture.

2.Teaching content

"Settlement reconstruction" module selects typical domestic and foreign settlement[s] as research objects, guides students to extract components of settlement spac[e] from the perspective of morphology and typology with the help of map software an[d] literature materials, and design an "ideal city" based on the prototype of settlemen[t] space. The teaching process lasts for four weeks (including one week of drawin[g] evaluation), including three exercises:

(1) One person in a group, complete the figure ground relation analysis of a typica[l] settlement form.

(2) Two persons in a group, aiming at a typical settlement space element, comple[te] narrative expression model of a settlement space sequence.

(3) Two persons in a group, basing on the prototype of settlement space, design a[n] "ideal city" within the provided framework.

3.Teaching process

The first week

Settlement space graph

(1) Lecture on Teaching of World Settlements.

(2) Select a typical settlement of specific type, draw a figure ground relation analys[is] diagram of the settlement with the help of map software and literature materials.

The second week

Settlement space element

(1) On the basis of settlement analysis diagram, aiming at the relationship betwee[n] a settlement space element and daily life, complete narrative expression model of settlement space sequence.

(2) Make architectural models that can express the spatial sequence of the settlement.

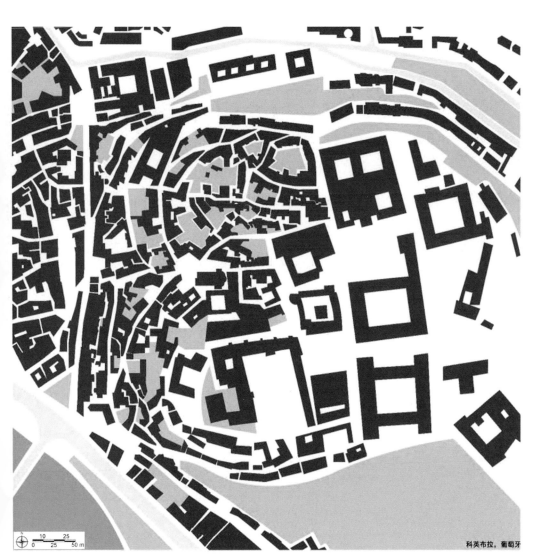

科英布拉，葡萄牙

科英布拉
Coimbra

建筑系统分析图

道路系统分析图

景观系统分析图

he third week

econstruction of settlement space (model)

) Basing on specific settlement space elements and prototypes, within a given nge (such as 100 cm × 50 cm × 50 cm), design an "ideal city" that meets the nagination of future life and an "invisible settlement".

) Detail the model and summarize the final results.

The fourth week

Defense.

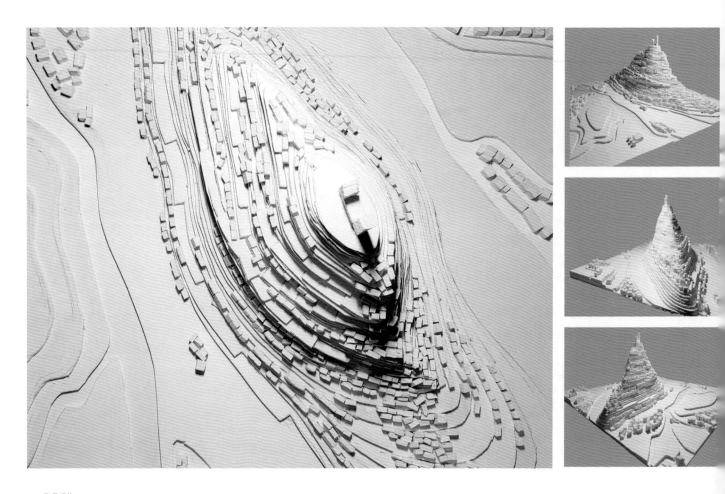

聚落重构

圣马力诺整座城市坐落于提塔诺山，其建筑物的分布与蜿蜒的山势完全吻合，总体呈现出等高线形的聚落特点。台阶是我们本次表达的重要空间元素，但这里的台阶不同于传统的阶梯，而是因山势的起伏而逐渐增高起伏的山路。在1:1000的模型中，我们用等高线的轮廓抽象表现地形，不仅设计了沿着山势环绕蜿蜒的山路、山上交错的道路交叉口，还对上山下山的道路进行了编排，进一步丰富了山的道路系统。圣马力诺最重要的建筑是教堂、宫殿以及位于山巅的城堡，我们由此抽象山最重要的建筑元素——峭壁城堡，并沿着山势建造出居民的住所、广场、教堂等一系列生活场所。我们选用了许多忽略自身造型的小方块来抽象表达出这些建筑物沿着地形分布的形态。

Settlement reconstruction

The whole city of San Marino is located on Mount Titano. The distribution of the buildings is completely consistent with the winding mountain, and the overall appearance shows the characteristics of contour linear settlement. The step is the important spatial element in the expression. However, the step here is different from the traditional step, but the rolling mountain path gradually increases with the ups and downs of the mountain. In the 1:1000 model, we use contour lines to abstract the terrain, not only design the winding mountain path and interlaced road intersections along the mountain, but also arrange the roads up and down the mountain to further enrich the road system of the mountain. San Marino's most important buildings are the church, the palace and the castle located on the top of the mountain. From this, we abstract the most important architectural element — the cliff castle, and build a series of living places such as residences, squares and churches along the mountain. We choose a number of small blocks that ignore their own shape to abstract the shape of the buildings along the terrain.

圣马力诺因其"易守难攻""防御型"地形地势，自14世纪以来就与外界隔离开来，如世外桃源一般。因此这里的居民，生活悠闲自在，与世隔绝，进行的空间活动也简单、类似。主要围绕着自己的住宅四周以及教堂展开，山上教堂林立，几乎每个社区居民都有独立的教堂，方便社区居民进行祷告。我们也因此在每一片小方块中，用较大的方块表现教堂，还将空出的纸板作为居民们聚会聊天的广场。

"理想之城"

设计重点：将水平聚落垂直，形成一座特征明显的"阶之城"

设计背景：由于人口持续增长，地球不足以为人类提供足够资源，我们必须在宇宙中寻找一颗宜居的星球，延续人类繁殖。而我们的这座理想之城就建立在这样一个星球之上。为了使人类能够在这座星球上持久生存，我们汲取地球上生活的经验教训，要最大限度地利用空间，于是我们将原本水平的聚落垂直，使得空间元素重新组合，增大利用效率。

设计理念：

1 这座星球重力消失，或者说重力接近于0，于是空间能被双重使用，一座房子、大厦的另一面仍是同样能够使用的空间，这就出现了双面建筑。模型中每一立方体都是一个能够双面使用、呈对称的建筑物。

2 为节约空间，我们省去了传统作为连接的台阶，而将房屋堆砌起来，形成供人们行走的"台阶"，节约空间的同时为人们提供便利。如图所示，模型中的呈"台阶"状的路径皆由房屋组成，在较为宽敞的平面还形成供人文赏玩、交谈的广场以及公园。

3 房屋由双面构成，行人均可以在上面行走，我们用纸片人的形式来张贴表达出这一特点，也因此当阳光直射顶面异常炎热时，人们可以在另一面进行空间活动，以达到最理想的效果，为人们带来最好的体验。

中国传统建筑文化
TRADITIONAL CHINESE ARCHITECTURE

赵辰 史文娟

建筑学专业学习的主课为建筑设计，与设计相关的众多学科知识（博雅），最终需要通过设计来实现（贯通）。故，对中国建筑史的学习，当既学中国传统建筑的形式语言，更需了解传统建筑形式背后的历史，思辨传统技术、生活、哲学、美学、制度等等与其时建筑形式之关联，"知其然，知其所以然"；将建筑历史的教学与建筑设计的教学更紧密地结合，成为建筑设计讲理念思考依据。最终，希望借课程的教与学之持续思考，在以西方文化背景的建筑学领域中，逐步构建中国文化价值的建筑学术话语体系。

以当代建筑学的理论框架，中国传统建筑文化被表述为三个领域："建构""人居""城镇"。"建构"者，在单体建筑的营造层面，以木构框架与土构围合为基本特征；"人居"者，在建筑群体的空间组织层面，以院落为核心；"城镇"者，在聚落空间与城市形态层面，由村落而至都城，以中国特色的城市发展规律为中心。课程设置"建构文化""人居文化""城镇文化"三大专题，成为系列讲座；每个专题系列，辅以相关设计作业，以实际的空间与图形操作来加深对专题内容的理解。

其中，"建构"专题，以"中国房子"（基本单元）为题进行传统营造过程模拟，提供慈城、剑川、闽东北、徽州四个地域的基本建造类型（三开间、二层）及归纳出的建筑基本单元平、立、剖图纸。学生分组选择地域类型，建立数字三维模型，模拟营造系统的全过程。以此了解传统建造材料与工序，在了解、掌握其建造规律的同时，理解地域性的单体建筑形式与实际建造体系之间的必然关系。

"人居"专题，设置"中国院子"（空间组合体）的设计。课程为学生提供江南某历史县城一街廊内的大小两种类型的基地及相关建筑要素模型（门屋、厢房、游廊、院墙等）。要求学生选定基地类型后，首先进行传统生活起居行为模式的"脚本"写作（目前设定以明清之际江南传统城市中某大户人家，如"李商人家""张员外家"等为主题），策划日常生活起居过程，并对相应的生活空间加以描述限定。随后，学生在所选基地上依据地界，以给定的建筑要素，根据前期自行设定的"脚本"，组合构建院落的三维模型体。经此过程，了解中国传统院落空间组合的基本设计原则和方法，理解其传统的生活起居行为之内容及其建筑类型与形式的规律。

以上课程调整历经2018年、2019年，从这两年的课堂效果与两次作业情况看来，

以专题讲座与作业形式结合的教学模式，可以最大限度地调动学生的学习积极性并拓展思维。尤其"中国院子"，学生被文本写作环节吸引，主动查找、阅读大量相关史料；在随后的"设计"环节，自然而然思考院落生活中的衣食起居，考虑脚本人物的行为流线、建筑功用，对脚本与设计前后做校对调整；中轴线之外的偏院、隙地，不约而同选择了园林设计，真实地反映了中国传统人居行为之规律。

中国传统建筑博大精深，如何在一学期的课程教学中，梳理脉络，提炼观点，如何给学生系统而完整的中国建筑历史，并经由史实而史论，再到设计实践，而不浮于知识点的掌握，以培养未来职业建筑师的理论能力，提供其创新的源泉，并激发学生研究传统建筑历史的兴趣，是今后教案改革需要不断探索的最重要目标。

The main course of architecture major is architectural design, and the knowledge of many subjects related to design (erudition) will be realized (comprehension) through design.Therefore, when studying the history of Chinese architecture, we should not only learn the formal language of traditional Chinese architecture, but also understand the history behind the traditional architectural form, speculate on the relationship between the traditional technology, life, philosophy, aesthetics, system, etc. and current architectural form, "learn firstly, and then how why"; combining the teaching of architectural history with the teaching of architectural design more closely, become the theoretical basis of architectural design. Finally, it is expected to gradually build an architectural discourse system of Chinese cultural value in the field of architecture with western cultural background through continuous thinking on the teaching and learning of the course.

According to the theoretical framework of contemporary architecture, Chinese traditional architectural culture is expressed in three fields: "construction",

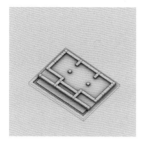  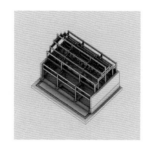 

"human settlement" and "town". "Construction", on the construction level of single buildings, takes the wooden frame and soil structure enclosure as the basic characteristics; "human settlement", on the spatial organization level of architectural group, takes courtyard as the core; "town", on the level of settlement space and urban form, from villages to capitals, centers on urban development rules with Chinese characteristics. Three major subjects of "construction culture", "human settlement culture" and "urban culture" have been set up as a series of lectures in the course; each series are supplemented by related design assignments, using actual spatial and graphical operation to deepen the understanding of the subject content.

In the subject of "construction", traditional construction process is simulated in the theme of "Chinese house" (basic unit); basic construction types (three rooms and two floors) in four regions including Cicheng, Jianchuan, southeast of Fujian, Huizhou are provided, and plan, facade and cross-section drawings of the basic building units are concluded. Students choose region type in groups, build digital 3D model, and simulate the whole process of construction system, so as to understand traditional construction materials and procedures, understand the inevitable relationship between the regional single architectural form and actual construction system while understanding and mastering the construction rules.

In the subject of "human settlement", design of "Chinese courtyard" (space combination) is set. In the course, a large foundation and a small foundation in a block of some historical county in the south of Yangtze River, and relevant architectural element models (gate house, wing room, veranda, courtyard wall, etc.) are provided to students. Students are required to write "script" in the mode of traditional living behavior after choosing foundation type (currently it is set as some wealthy family in traditional city in the south of Yangtze River in Ming and Qing Dynasties, such as the theme of "family of merchant Li" "family of landlord

Zhang"), plan process of daily life, and describe and define relevant living space. Later, students are required to combine and construct 3D model of courtyard on the selected foundation with provided architectural elements according to the previously independently set "script". In this process, students can understand the basic design principles and methods of Chinese traditional courtyard space combination, and the content of traditional living behavior and rules of architectural type and form.

Aforesaid course has been adjusted in 2018 and 2019, the class effect in these two years and situation of two assignments show that the teaching form that combines special lectures and assignments can arouse student's study enthusiasm and expand thinking to the greatest extent. Especially in "Chinese courtyard", students are attracted by the writing, and actively look up and read a lot of related historical materials; in the subsequent "design", they naturally think about the daily life in the courtyard, consider the behavior line of the script character and architectural functions, proofread and adjust the script and design; for the side courtyard and clearance out of the axis, they all choose garden design, which truly reflects the rules of Chinese traditional human settlement behavior.

Chinese traditional architecture is extensive and profound, how to rationalize and refine opinions in the course teaching of a semester, how to provide students with systemic and complete Chinese architectural history, through the historical facts and theory, then to design practice, instead of floating on the mastery of knowledge, so as to cultivate the theoretical ability of future architect, provide the source of innovation, and stimulate students' interest in the study of traditional architecture history, is the most important objective for constant exploration of teaching reform in the future.

"中国房子"（基本单元）的营造设计

1.教学目标

了解中国传统建造体系的基本单元的建造材料、建造程序，理解并掌握其建造的规律，乃至可适应不同地域的建筑形式。

2.教学方法

数字三维模型，模拟营造系统。

3.教学进程

第一阶段，根据提供的"中国房子"基本单元之平、立、剖图纸的数据信息以及各个地域性建造体系的信息，建立"中国房子"的四种案例模型。

      （1）慈城

      （2）剑川

      （3）闽东北

      （4）徽州

第二阶段，建立"中国房子"模型，根据体形改变的要求进行调整。

进深：步架数，屋面形体变化（屋檐、屋脊）

通面阔：开间尺寸

尺度：

      面阔：3开间；10～12 m

      进深：5～6步架；步架单位尺度根据地域规律

      高度：1～2层，底层3～4 m，二层3 m（檐口高度）

基础与台基：夯土地基处理

      砖石台基；踏步、栏杆；铺地（砖、石）

      基础：墙体基础（墙基）；构架基础（柱础）

构架：

      木构屋架：进深方向的柱、梁枋构成之各缝屋架；

                 面阔方向的梁枋、檩（桁）条联系各缝屋架

      屋面：木构椽子；屋面板

围合：

      墙体（夯土、砖石）

      各部位墙体：山墙、檐墙、院墙

      开启：门洞（门头）、窗洞

屋面：

      硬山：山墙的各种变体（地域差异）和构造（排水）

      悬山：木构的出际（檩条从山墙的悬出）

      瓦作：仰瓦、底瓦；勾头滴水

      屋脊（各种屋脊）：正脊、戗脊、垂脊

Construction Design of "Chinese House" (Basic Unit)

1.Training objective

To know the construction material, construction procedure of basic unit of Chinese traditional construction system, understand and master the rules of construction, method and even adapt to the architectural forms in different regions.

2.Teaching method

Digital 3D model, simulation construction system

3.Teaching process

Stage 1: according to the data information of plan, facade and cross-section drawing of the basic unit of "Chinese house" and information of regional construction system, establish four case models of "Chinese house".

      (1) Cicheng

      (2) Jianchuan

      (3) Northeast of Fujian

      (4) Huizhou

Stage 2: adjust the established "Chinese house" model according to the requirement on size change.

      Access depth: horizontal spacing between purlins, roof form change (eaves, ridge)

      Total width: room dimension

      Dimension:

            Width: 3 rooms; 10~12 m

            Depth: 5~6 horizontal spacing between purlins; according to the regional rule

            Height: 1~2 floors, ground floor 3~4 m, second floor 3 m (cornice height)

      Foundation and stylobate:

            Rammed foundation treatment

            Brick stone stylobate; step, rail; pavement (brick, stone)

            Foundation: wall foundation (wall foundation); structure foundation (column foundation)

      Structure:

            Wooden roof truss: column, beam in depth direction compose roof truss;

            Beam, purlin in width direction link the roof truss

            Roof: wooden rafter; roof panel

      Enclosure:

            Wall (rammed earth, masonry)

            Wall of each part: gable wall, eave wall, courtyard wall

            Opening: door opening (door head), window opening

      Roof:

            Flush gable roof: variations of gable wall (regional difference) and structure (drainage)

            Overhanging gable roof: overhanging of wooden structure (purlin is overhung from gable wall)

            Tilework: tilting tile, bottom tile; eave tile drip

            Ridge (various ridges): main ridge, diagonal ridge, vertical ridge

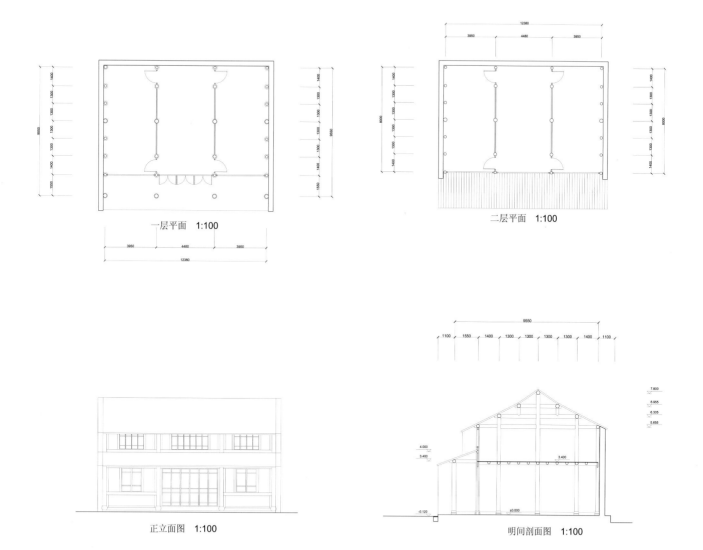

一层平面 1:100

二层平面 1:100

正立面图 1:100

明间剖面图 1:100

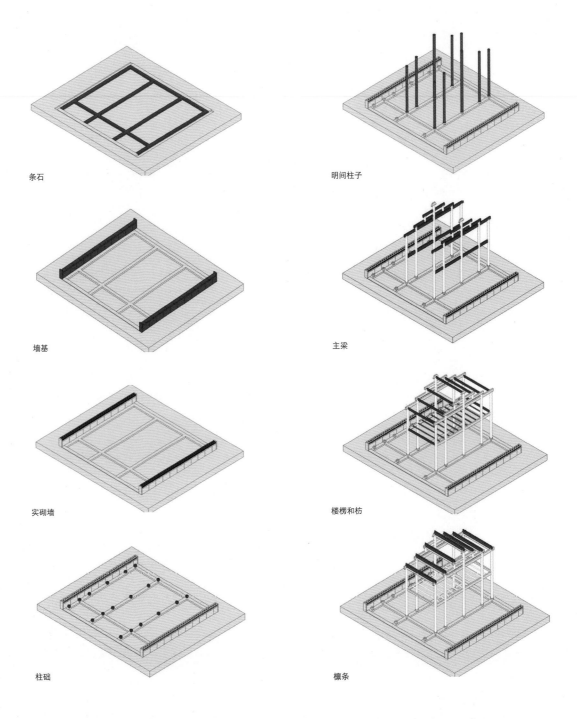

条石

明间柱子

墙基

主梁

实砌墙

楼楞和枋

柱础

檩条

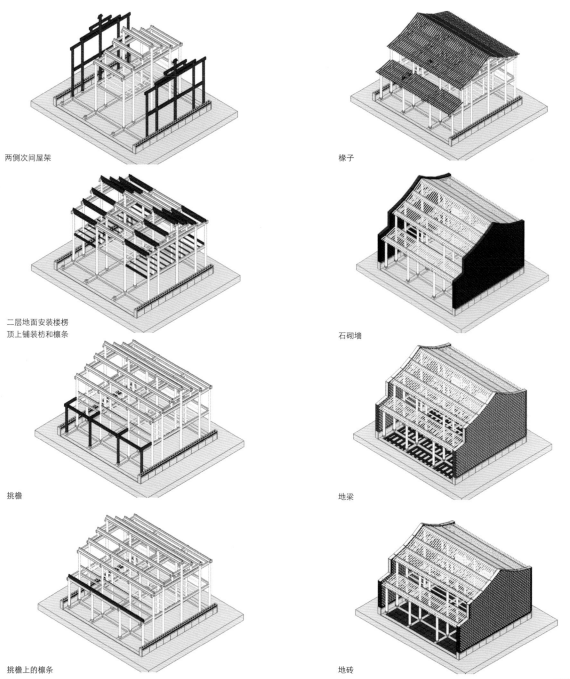

两侧次间屋架

橡子

二层地面安装楼楞
顶上铺装枋和檩条

石砌墙

挑檐

地梁

挑檐上的檩条

地砖

"中国院子"（基本组合体）的设计

1.教学目标

了解中国传统院落组合的基本建筑群体的规划设计原则和方法，理解并掌握其建造的规律，乃至可适应不同内容的建筑类型与形式。

2.教学方法

数字三维模型。

3.教学内容

按"中国院子"的基本建筑组合体的规律，规划设计一中、大型中国传统院落组合建筑群，以大户人家生活起居为行为模式，但也可能变形为祠堂、衙署、寺庙等。

4.教学进程

第一阶段：脚本策划（第一周完成）

通过中国传统城镇的人居生活的设想及描述，进行空间内容策划，为第二阶段"中国院子"的空间模型设计制作提出文字脚本。

规模选择：

(1) 大型住家，占地面积1500~2500 m²，出入口3~4个（2位同学合作完成）。

(2) 小型住家，占地面积500~800 m²，出入口2~3个（1位同学独立完成）。

写作方法：

分别以"张员外家（大型）""李商人家（小型）"为题，策划一明清时期江南传统家庭生活起居过程，需要设想人物、日常生活以及亲朋来访、节庆活动等，对相应的生活空间与场景进行描写创作，并归纳出"活动清单"。

第二阶段：空间模型（第二、三周完成）

根据给定的地界/地块的边界，以"中国房子"的基本单元、厢房、游廊、院墙、门屋等作为构成元素，以院落为中心组合成"中国院子"的基本建筑组合体。

尺度：

通面阔：路；中路、东西路；中路面阔为主，边路减之。

总进深：多进（两进以上）院落为单元，纵向发展；每一进单体的进深以主体建筑为主，步架数和进深尺度为最，其余随之。

院落：

开敞空间的院子：因所处区位尺度、形式、材料而定义。

开敞空间院落的变体：庭院；天井；偏院；园子。

主体建筑：相应院落为主体。

围墙、门屋、厢房等成为院落空间的围合边界。

配房：

厢房：与正房相配合纵向发展。

灶间、工房、杂院。

围墙：

作为院落组合体的主要围合元素，与其他建筑单体共同构成院落形态。

院墙：墙基、墙身、墙顶（瓦作），根据所处位置而形式、高度不同。

工程要点：端部、转角；高差。

入口：

主入口（墙门）：礼仪性出入口，形势变化较多，尺度较大，双开门。

后门：主入口的背向，生活性出入口，结合灶间、工房、杂院。

边门（便门）：侧门，结合廊道。

地面：

廊道：有屋檐覆盖的空间，与屋檐外有高差，按室内铺地做法。

阶沿：防水要求，强度要求。

院子：排水要求，防滑要求，耐久性。

Design of "Chinese Courtyard" (Basic Complexity)

1.Tranining objective

To know the planning and design principles and methods of traditional Chinese courtyard combination, understand and master the construction rules and even adapt to different architectural types and forms of different contents.

2.Teaching method

Digital 3D model.

3.Teaching content

According to the rule of the basic building combination of "Chinese courtyard", plan and design a medium or large traditional Chinese courtyard combination building group, with the life of wealthy family as the behavior pattern, or may be transformed into ancestral hall, government office, temple, etc..

4.Teaching process

Stage 1: script planning (completed in the first week)

Base on the imagination and description of the life in traditional Chinese towns, plan space content, propose script for the design and production of space model of "Chinese courtyard" at stage 2.

Scale:

(1) Large residence, with land area of 1500~2500 m², 3~4 accesses(2 students work together).

(2) Small residence, with land area of 500~800 m², 2~3 accesses(1 student completes independently).

Writing method:

With the title of "Landlord Zhang's family (large)" and "Merchant Li's family (small)" respectively, plan the living process of traditional family in the south of Yangtze River in Ming and Qing Dynasties, conceive characters, daily life, as well as visit from relatives and friends, festival celebration activity, etc., describe and create corresponding living space and scene, and summarize the "activity list".

Stage 2: space model (completed in the second, third week)

According to the boundary of the given land/plot, with the basic unit, wing room, veranda, courtyard wall, etc. of "Chinese house" as the elements, with the courtyard as the center, compose the basic architectural combination of "Chinese courtyard".

Dimension:

Total width; road, middle road, east and west road; middle road width mainly, side road is accordingly reduced.

Total depth: with multi-access (more than two) courtyard as the unit, longitudinal development; the depth of every access unit is subject to main building, horizontal spacing between purlins and access depth are the maximum; followed by the rest.

Courtyard:

Courtyard of open space: defined by location, scale, form, material.

Variation of courtyard of open space: courtyard; patio; side courtyard; garden.

Main building: relevant courtyard as the main body.

Fence, gate house, wing room, become the enclosing boundary of the courtyard space.

Auxiliary house:

Wing room: vertical development coordinated with the main room.

Kitchen, workhouse and courtyard.

Enclosure wall: as the main enclosure element of the courtyard assembly, it forms the courtyard shape together with other individual building.

Courtyard wall: wall base, wall body, wall top (tilework). According to the location, the form and height are different.

Key points of the project: end, corner; altitude difference.

Entrance:

Main entrance (wall gate): ceremonial entrance, with more situation changes, larger size, double door.

Back door: back of the main entrance, living entrance, combined with kitchen, workshop, utility yard.

Ground:

Corridor: space covered by eaves, has height difference with the outside of eaves, according to the practice of indoor paving.

Step edge: waterproof requirement, strength requirement.

Courtyard: drainage requirement, slip resistance requirement, durability.

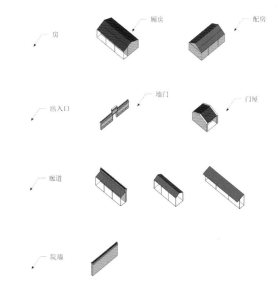

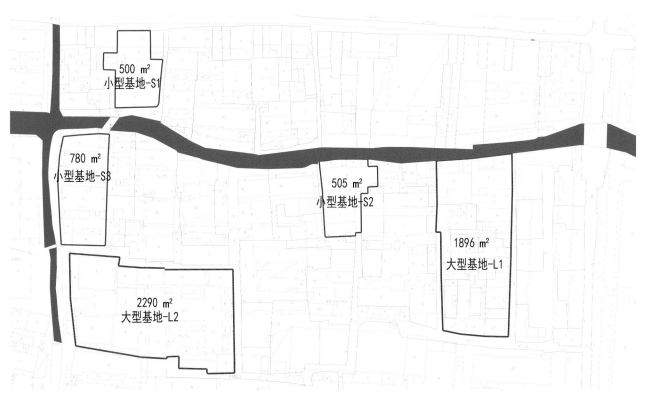

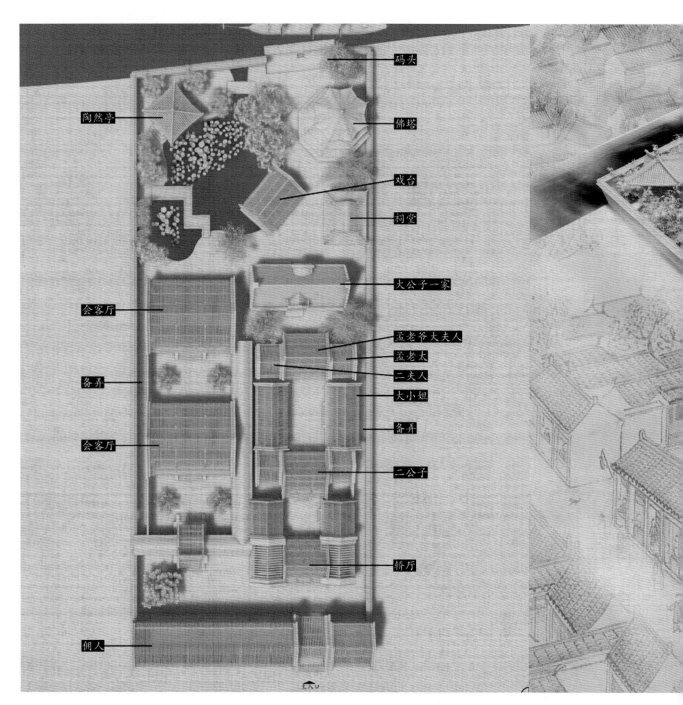

码头

佛塔

戏台

祠堂

陶然亭

大公子一家

会客厅

孟老爷大夫人

孟老太

二夫人

备弄

大小姐

备弄

会客厅

二公子

轿厅

佣人

26

大学生健身中心改扩建设计
RECONSTRUCTION AND EXPANSION DESIGN OF COLLEGE STUDENT FITNESS CENTER

傅筱 王铠 钟华颖

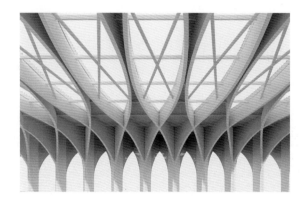

1.课程缘起

大学生健身中心改扩建是南京大学建筑本科新开设的一个三年级设计课程。在南京大学过去的设计教学中一直缺少一个专项训练中大跨结构的设计课程，而是将其融入三年级商业综合体课程设计中。学生既要解决商业综合体的复杂城市关系和复合功能空间，又要研究其中的大跨度设计，其难度不难想象。大学生健身中心改扩建设计课程开设的初衷十分明确，即是将中大跨空间从商业综合体中分离出来进行分项训练。

2.选题思考：弱规模，强认知

在中大跨度建筑设计训练方面国内许多建筑院校做出了值得学习的探索，它们多数是以中大型体育馆、演艺中心或者文化中心为授课载体，以结构选型为基础，结合设备训练学生对大型建筑场馆的综合设计能力。有的院校将结构独立成8周专项训练，然后再结合进8周综合馆设计中。这些训练通常是放置在四年级或者毕业设计，不少院校设计周期长达16周，让学生得到较为充分的训练。对于南京大学"2+2+2"的建筑教育模式而言，是不适宜进行长周期训练的。在短短的8周教学时间内，我们应该让学生建立何种认知和培养何种能力是关键。教学团队意识到高校课程设计并非实际工程训练，与漫长的职业生涯相比，学生的认知能力比实际操作能力更为重要，设计的规模大和技术的复杂性并不一定能提升他们对设计本质的理解。因此，结合南京大学自身的教学特征，教学团队在选题上做了如下思考。

首先，缩小建筑规模，弱化功能配比。其意义在于既适合于短周期训练，又可以减轻学生在设计过程中的绘图、模型制作工作量，从而将教学重点放置在设计思维的探讨上。在教学过程中，建议学生少做表现性实体模型，如果电脑三维技术娴熟，甚至可以抛弃实体模型推动设计的传统方法，只是在早期概念阶段辅助以模型研究。

其次，将结构训练的重点放在结构与形态、空间的关联性上。在以前的课程设计中学生较为熟悉的是框架、砖混结构，虽然每一次课程设计均与结构不可分割，但是结构被外围护材料包裹，在建筑表达上始终处于被动"配合"的状态。对于有一定跨度的空间，容易让学生建立一种"主动"运用结构的意识。为了达到这个训练目标，教学团队认为跨度不宜太大，结构选型余地较大，结构与空间的配合受到技术条件的制约相对较小，有利于学生充分理解空间语言与结构的关系。

第三，强调建筑设计训练的综合性，结构只是一个重要的要素。即不过分放大结构的作用，在将结构作为空间生成的一种推动力的同时要求学生综合考虑场地、使用、空间感受、采光通风等基本要素。实际上通过短短8周教学就希望学生能够达到实

际操作层面的综合性是不现实的，教学的目标是让学生建立起综合性的认知。所以题目设置宜简化使用功能，明确物理性能要求，给定设备和辅助空间要求，将琐碎的知识点化为知识模块给定学生。学生更多的是学会模块间的组织，而不需要确切了解其中的每一个技术细节。技术细节将随着学生今后的职业生涯而不断增长，技术细节在今天确实日新月异，在高校教授过多的技术细节或者教师的所谓工程经验，也许学生毕业时这些细节和经验就已经面临淘汰。所以在高校应教授"认知"，进而教授由正确认知带来的技术方法，最终升华为学生的设计哲学。因此，在具体的教学中，结构选型是指定的，但鼓励学生结合性能需求进行合理改变；采光通风要求是明确的，但鼓励学生结合人的需求进行设计；设备和辅助空间是给定的，但要求学生学会在空间上进行合理布置，理解只有设备和辅助空间布置妥当才能创造使用空间的价值。

3.教学方法：从"一对一"到"一对多"

本科建筑设计教学方法通常是采用教师一对一改图模式，这种模式的优点是教师易于将设计建议传递给学生，当学生不理解时，还可以方便地用草图演示给学生。这适合于手绘时代的交流，甚至教师帅气的草图也是激励学生进步的一种手段。其缺点是学生的问题和教师的建议均局限在一对一的情景中，不能让更多的学生受益，如果教师发现共性问题须临时召集学生再行讲解，其时效性、生动性都大打折扣。鉴于此，我们采用了"一对多"的改图模式，教师和学生全部围坐在投影仪前面，由每个学生讲解自己的设计，教师做点评，并同时鼓励其他学生发表点评意见，当教师认为需要辅助草图说明问题时，教师采用Pad绘制草图，并让每个学生都能看见。

从教学效果看，一对多的改图模式的优点是明显的。首先这样的改图方式具有较好的课堂氛围，学生面对大屏幕讲解自己的设计增强了课堂的仪式感，仪式感的背后是对学习的敬畏之心，同时因为教师不是站在高高的讲台上，又具有平等讨论的轻松气氛。从教学效果上看，学生既体验到当众讲解的压力，也体验到因出色的设计呈现带来的喜悦，因而无形中激发了他们用功的积极性。其次有利于教师发现学生的共性问题，避免同一问题反复讲却反复犯的教学通病，大大提高了课堂效率。

1.The background of course

The reconstruction and expansion of the college student fitness center is a new third-year design course of the School of Architecture and Urban Planning, Nanjing University. In the past design teaching of Nanjing University, there was always sho

of a special training course of long-span structure design, which was integrated into the course design of the third-year commercial complex. The students should not only solve the complex urban relationship and composite function space of commercial complex, but also study the long-span design, the difficulty of the course is not hard to imagine. The original intention of setting the design course of the college student fitness center reconstruction and expansion is very clear, that is, to separate the long-span space from the commercial complex for itemized training.

.Topic thinking: weaken scale, strength cognition

In terms of architectural design training of long-span, many architectural colleges in China have made the exploration which is worth learning. Most of them took medium and large gymnasiums, performance centers or cultural centers as teaching carriers, and trained students' comprehensive design ability for large construction venues based on structure selection and combined with equipment. Some colleges separate the structure into 8 weeks of special training, and then combine it into 8 weeks of comprehensive venue design. These trainings are usually placed in the fourth grade or graduation project, the design cycle of many colleges up to 16 weeks, so that the students could get adequate trainings. As for the "2+2+2" architectural education mode of Nanjing University, it is not suitable for long-periodic training. In just eight weeks of teaching, the key is to help students to establish some kind of cognition and develop some kind of ability. The teaching team realized that college curriculum design was not practical engineering training, the students' cognitive ability was more important than their practical operating ability compared with their long career, and the scale and complexity of design did not necessarily improve their understanding of the nature of design. Therefore, combined with Nanjing University's own teaching characteristics, the teaching team made the following considerations on the topic selection.

Firstly, it should reduce the building scale and weaken the function ratio. The significance lies in that it is not only suit for short period training, but also lessen the workload of drawing and model making of students in the teaching process, so that the teaching emphasis can be placed on the discussion of design thinking. In

the teaching process, it is suggested that students should do less expressive entity model. If they are skilled in three-dimensional technology, they can even abandon the traditional method of entity models drive design and only assist the model research in the early conceptual stage.

Secondly, the emphasis of structural training is on the relationship between structure, form and space. In the previous course design, students are familiar with the frame and brick-concrete structure. Although each course design is inseparable from the structure, the structure is wrapped by the peripheral materials and is always in a passive state of "cooperation" in architectural expression. For a space with a certain span, it is easy for students to establish a sense of "active" using the structure. In order to achieve this training goal, the teaching team believes that the span should not be too large, about 30m is appropriate, and there is a large scope of structure selection, and the coordination between structure and space is relatively limited by technical conditions, so as to facilitate students to fully understand the correlation between spatial language and structure.

Thirdly, the emphasis is on the integration of architectural design training, the structure is only an important element. That is to say, the function of the structure is not exaggerated. While taking the structure as a driving force for the generation of space, students are required to comprehensively consider the basic elements of site, use, space perception, lighting and ventilation. In fact, it is unrealistic to expect students to achieve practical comprehensiveness in just eight weeks' teaching. The teaching goal is to enable students establish a comprehensive cognition, so the topic setting should simplify the use function, clarify the physical performance requirements, set equipment and auxiliary space requirements, and turn trivial knowledge points into knowledge modules for students. Then, students should learn more about the organization of modules, and do not need to know each of these technical details. Technical details will continue to grow with the students' future career, technical details in today is really changing with each passing day. Many technical details or teachers' so-called engineering experience are taught in colleges and universities, which may be obsolete by the time they graduate. Therefore, "cognition" should be taught in colleges and universities, and then the technical methods brought by correct cognition should be taught, which finally developed design philosophy of students. Hence, in specific teaching, structure selection is specified, but students are encouraged to make reasonable changes based on performance requirements; the requirements for lighting and ventilation are clear, but students are encouraged to design according to human needs; equipment and auxiliary space are given, but students are required to learn to arrange the space reasonably, and understand that only when the equipment and auxiliary space are properly arranged can the value of using space be created.

3.Teaching method: from "one-to-one" to "one-to-many"

The undergraduate architectural design teaching method usually adopts the mode of one-to-one figure modification by teachers, the advantage of this model is that teachers can easily pass design suggestions to students, and when students don't understand, they can also easily demonstrate to students with sketches, which is suitable for communication in the hand-drawn era. Even the handsome sketches of teachers are means to encourage students' progress. And the disadvantage is that students' problems and teachers' suggestions are limited to one-to-one situation, which cannot benefit more students. If teachers find common problems and need to call on students to explain them temporarily, the timeliness and vividness will be greatly reduced. In view of this, we adopted "one to many" figure modification mode, all teachers and students sit in front of the projector, then, each student explains his own design, the teacher makes comments, and encourages other students to give comments. When the teacher thinks it is necessary to help the sketch to explain the problem, the teacher uses Pad to make the sketch, which can be seen by every student.

From the perspective of teaching effect, the advantages of one-to-many figure modification mode are obvious. First of all, this kind of figure modification way has a good classroom atmosphere. Students explain their design in front of the big screen, which enhances the sense of ceremony in class. Behind the sense of ceremony is the fear of learning, and because the teacher is not standing on the high platform, there is a relaxed atmosphere of equal discussion. From the perspective of teaching effect, students not only experienced the pressure of public explanation, but also experienced the joy brought by excellent design presentation, thus stimulating their enthusiasm to work hard. Secondly, it is beneficial for teachers to find out the common problems of students and avoid the common teaching mistakes of repeating the same problem, which greatly improves the classroom efficiency.

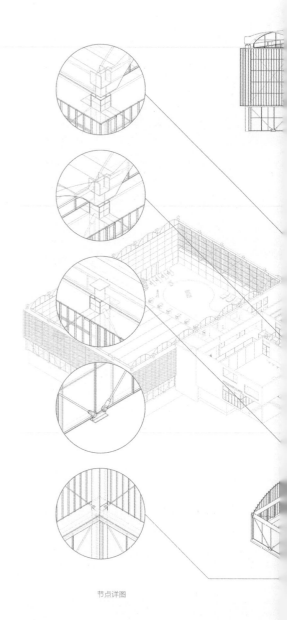

节点详图

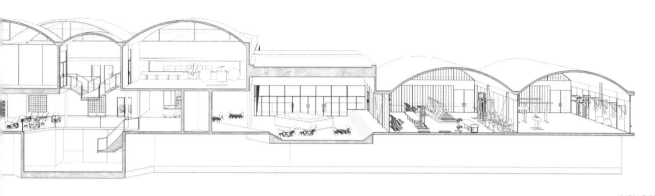

空间剖透视图

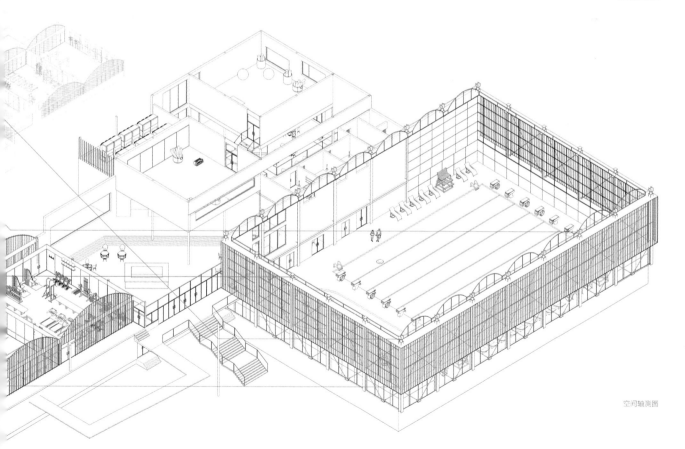

空间轴测图

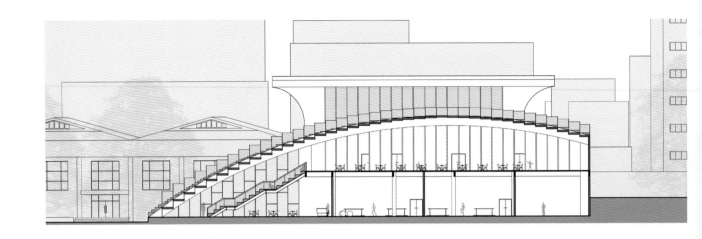

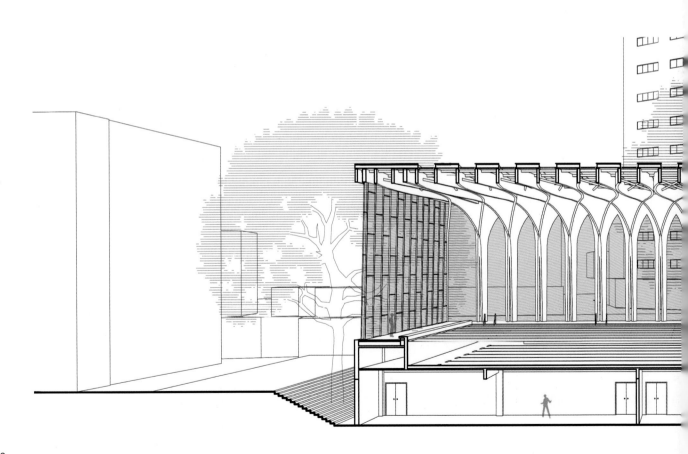

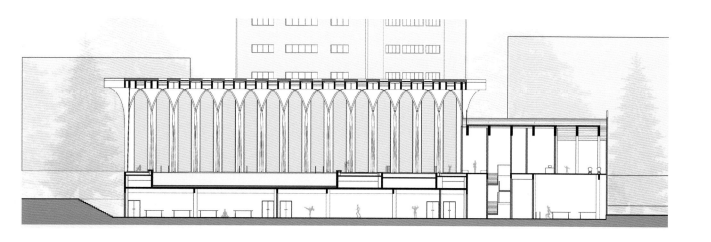

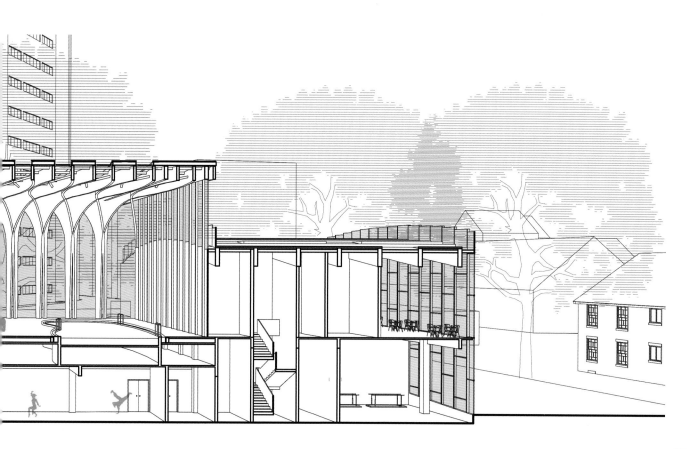

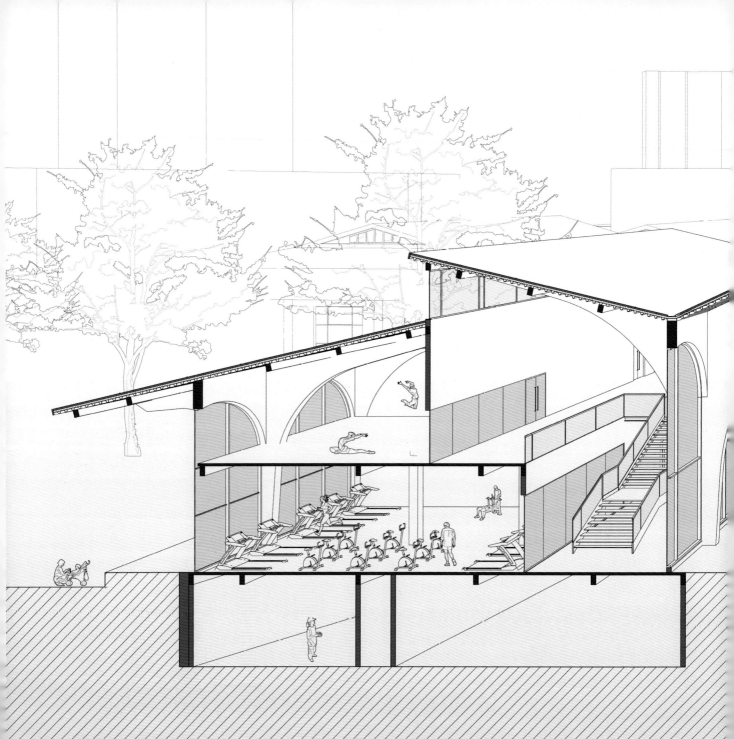

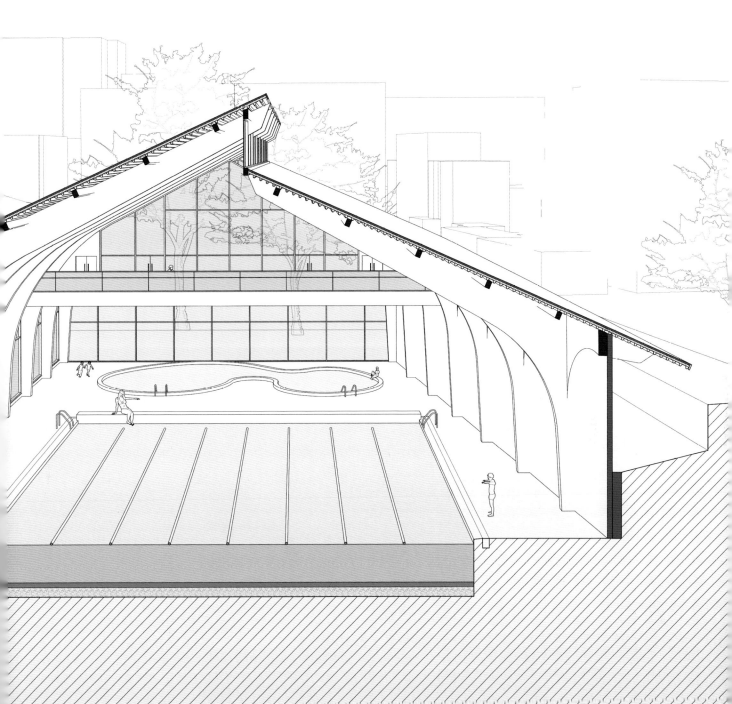

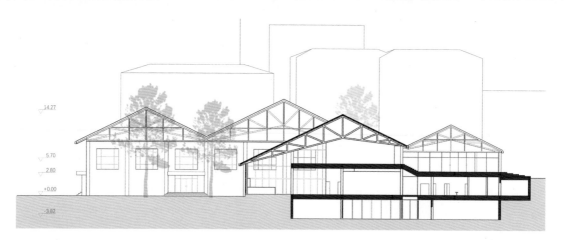

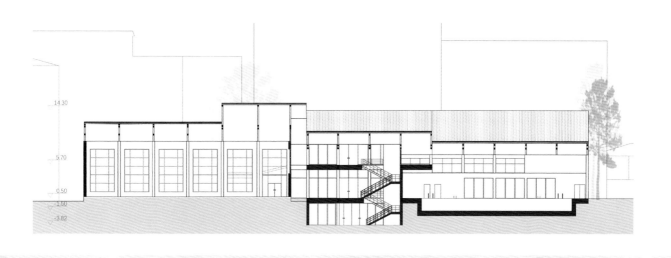

社区文化艺术中心
COMMUNITY CULTURE AND ART CENTER
王铠 钟华颖 尹航

1.教学目标

本课题以社区文化艺术中心为训练载体，学习并掌握综合功能建筑基本设计原理，理解老城区街区式建筑与城市环境的辩证关系，培养空间、建造与功能的综合组织能力。

2.城市建筑

设计题目选址于南京核心区域的一个密集建成区——百子亭风貌区一处旧城改造待建地块。面对过去三十年城市建成环境的剧变，这个地块集合了城市的历史、市井生活以及当代老城核心区再发展的综合命题。作为城市历史地段和危旧房改造的混合地带，社区文化艺术中心的功能主题，同城市更新对公共空间活力的激发，共同构成时代需求的一种趋向。同时题目恰当的规模和复杂性，使其与高年级建筑设计综合训练教学目标相符合。

3.公共性

我们的时代是公共领域极大化，同时个体对其参与和体验极其微弱的矛盾状态。我们可能用林奇的理论去规划建设一座新城，让它看起来很有"归属感"，却很难让住在一栋住宅里的居民拥有一个天天相遇、相识的场所。我们的城市拥有最宏伟的广场，而庆典之后我们还是要回到自己封闭的居所，不知道左邻右里为谁。基于这种思考，在课程设计教学中有意地强化了"公共空间"主题，试图激活"公共性"和归属感的生活体验。延续传统城市肌理的院落、围合、中心，这些可能引发公共性的体验概念，以各种可能进行尝试。或者是被放大的建筑内部交通空间来延续城市空间体验，唤醒城市活力与建筑功能空间的相关性。

4.课题内容

地上总建筑面积8000 m²，包含文化中心、演艺中心、社区商业以及必要的附属建筑空间和周边环境设计。

1.Training objective

This project takes the community culture and art center as the training carrier, aims to learn and master the basic design principles of comprehensive functional architecture, understand the dialectical relationship between the block architecture in the old city and the urban environment, and cultivate the comprehensive organizational ability of space, construction and function.

2.Urban architecture

The design topic is selected in a dense built-up area in the core area of Nanjing -- plot to be built for old city reconstruction in Baiziting area. In the face of the drama changes in the urban built environment over the past 30 years, this plot integrate the history of the city, urban life and the comprehensive proposition of redevelopme of the core area of contemporary old city. As a mixed zone of the historical area city and the renovation of dilapidated houses, the functional theme of communi culture and art center, together with the stimulation of vitality of the public space urban renewal, constitute a trend of the times. At the same time, the appropria size and complexity of the subject make it consistent with the teaching objective comprehensive training in senior architectural design.

3.Publicity

Our era is in the contradiction in which the public area is extremely large while th individual participation and experience are extremely weak. We may use Lync theory to plan and build a new city, so that it seems to have a "sense of belonging but it is difficult for the residents living in a house to have a place where they me and know each other every day. Our city has the grandest square, and after th celebration we still have to go back to our closed house, not knowing who's next us. Based on this thinking, the theme of "public space" is intentionally strengthened the course design and teaching, trying to activate the life experience of "publicity" a sense of belonging. Courtyards, encloses and centers of the traditional urban textu are continued, which may trigger the concept of public experience, and vario possible tries. Or the enlarged traffic space inside the building can continue the urb space experience and awaken the correlation between the urban vitality and t functional space of building.

4.Subject content

The total aboveground building area is 8000 m², including cultural center, performi art center, community business and necessary ancillary building space a surrounding environment design.

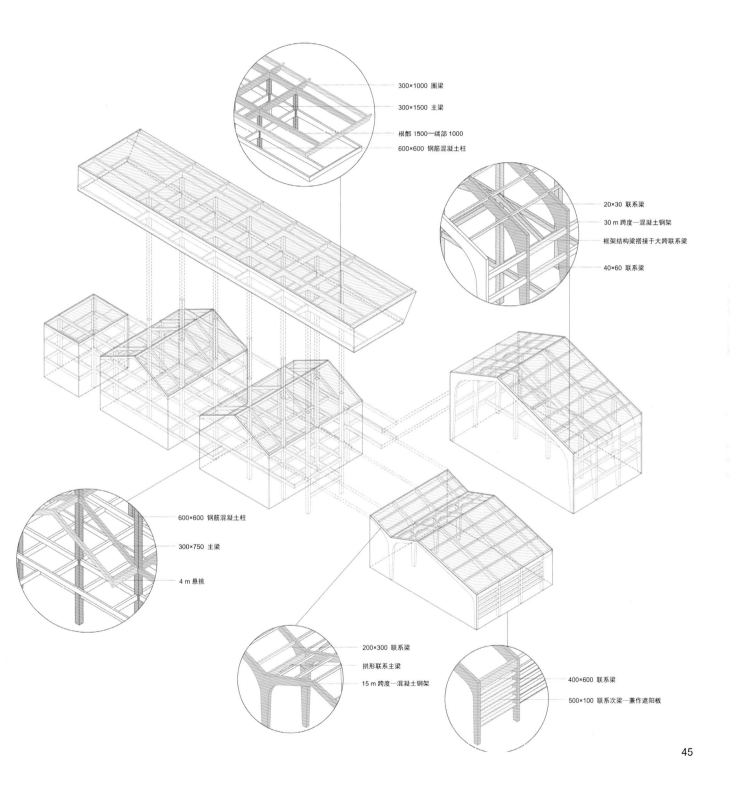

300×1000 圈梁

300×1500 主梁

根部 1500—端部 1000

600×600 钢筋混凝土柱

20×30 联系梁

30 m 跨度—混凝土钢架

框架结构梁搭接于大跨联系梁

40×60 联系梁

600×600 钢筋混凝土柱

300×750 主梁

4 m 悬挑

200×300 联系梁

拱形联系主梁

15 m 跨度—混凝土钢架

400×600 联系梁

500×100 联系次梁—兼作遮阳板

45

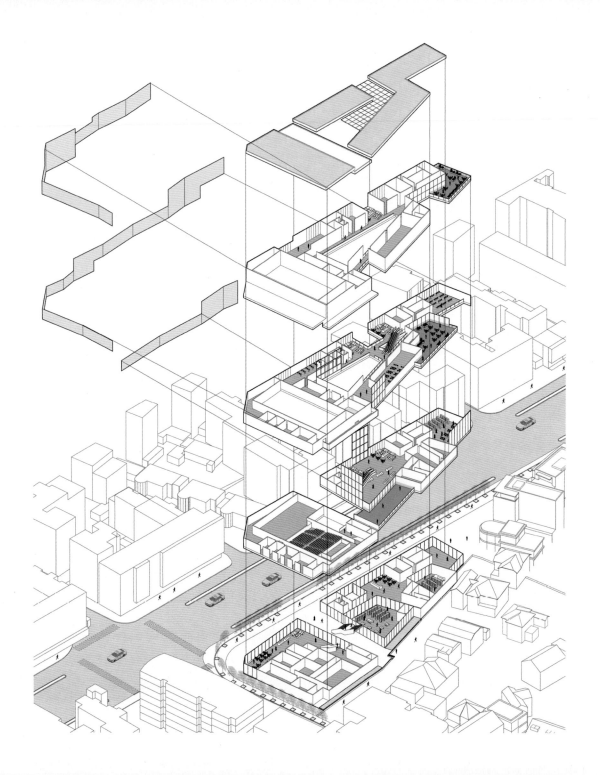

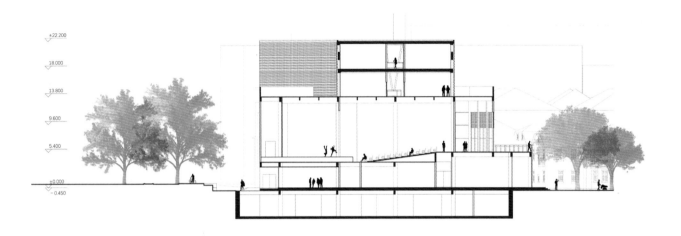

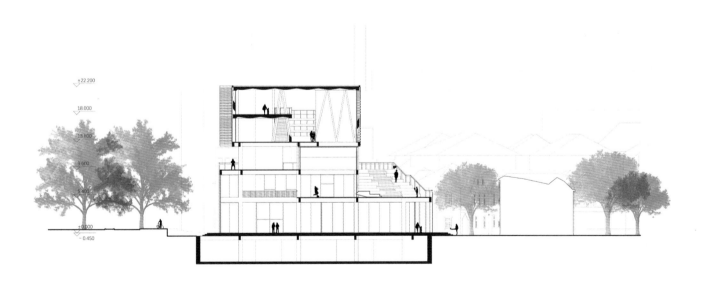

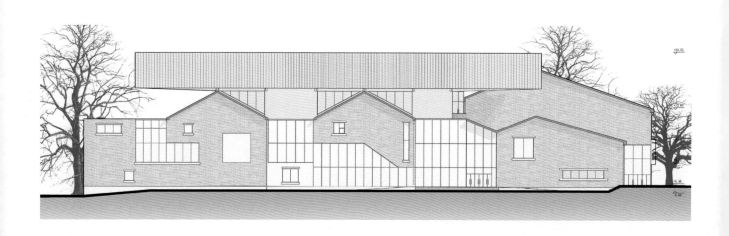

城市设计：密度与质量
URBAN DESIGN: DENSITY VS QUALITY

丁沃沃 唐莲 尤伟

19 世纪中叶之后，欧洲大陆在工业化的驱动下，经济社会蓬勃发展。其后短短的50年席卷欧洲的城市化改变了欧洲城市的结构和空间，改变了人们对城市的定义和认知。人们还没能足够地享受城市化带来的巨大经济效益，就不得不开始处理城市化带来的城市空间的问题。20 世纪以来，霍华德、赖特、柯布西耶提出了各种各样的城市形态模型，垂直城市的空间图景已经出现在柯布西耶的光明城市的主张中。之后，尽管人们批判了功能城市的非人文性，然而对垂直城市的探索却更加迫切：尤纳·弗里德曼的空间城市，矶崎新的聚合体和康斯坦的新巴比伦无一例外地针对城市扩张的压力，指出了基于既有城市规模，通过空间形态的创新，提高城市的承载能力是城市发展的最终出路。自20世纪中叶至今人们一直没有停止对可持续的城市形态这个问题的思考，探索还在继续。

经历了30多年快速城市化，我国沿海发达城市终于不得不结束轻松的扩张，开始面对土地资源终会枯等方式来提高土地使用效率，既是我们城市设计相关学术探索的重要方向，也是我们基于教研平台持续进行的系列教学实验之一。在此背景下，本课程基于真实地块的设计训练，认知高密度城市空间形态的真正含义，了解城市建筑角色和城市物质空间的本质和效能，初步掌握高质量城市空间和城市建筑组合之间的关系，同时，进一步深化空间设计的技能、方法与绘图能力。

对于第一次接触到城市设计的大四学生来说，在设计开展之前，首先需要建立起对城市空间形态的准确理解，为此城市空间形态的认知训练与设计训练同等重要；另外，城市设计图示不同于常规的建筑学图示，也是训练的核心内容。因此，在历时8周的教学设定中，围绕密度分析与场地调研的认知练习占时2周，密度/功能分配与深化设计占时4～5周，集中图纸表现占1～2周，绘图训练则贯穿了整个教学过程中。

After the middle of the 19th century, the European continent, driven by industrialization, achieved vigorous economic and social development. In the following 50 years, urbanization that swept through Europe changed the structure and space of European cities, as well as people's definition and cognition of cities. Before people can enjoy the huge economic benefits brought by urbanization, they have to deal with the problems of urban space brought by urbanization. Since the 20th century, Howard, Wright and Corbusier have put forward various urban form models, and the spatial prospect of vertical city has appeared in the bright city proposed by Corbusier. Later, despite people criticize the non-humanism of functional cities, th exploration of vertical cities becomes more urgent: spatial cities of Yona Friedma clusters of Arata Isozaki and new Babylon of Constant all point out that improveme of the carrying capacity of cities through innovation of space form on the basis existing city scale is the ultimate way out for city development. Since the middle of t 20th century, people have not stopped thinking about the sustainable urban form, ar the exploration still continues.

After more than 30 years of rapid urbanization, the developed coastal cities in Chir have finally had to end their easy expansion and begun to face the fact that lar will eventually be exhausted. Based on the city we live in, discussing to impro land use efficiency by changing land use nature, increasing building density ar improving transportation organization is not only an important direction of academ exploration related to urban design, but also one of the series of teaching experimer we continue on the teaching and research platform. Under this background, th course bases on real plot design training, aims to make students understand the re meaning of high-density urban space form, know the essence and effect of the role city building and urban physical space, preliminarily master the relationship betwee high quality urban space and urban architecture combination, at the same tim further deepen skill, method and drawing ability in space design .

To the senior students who contact urban design for the first time, they should fi establish an accurate understanding of urban spatial form before starting desig Therefore, the cognitive training of urban spatial form is equally important as t design training. In addition, urban design diagrams are different from convention architectural diagrams, which is also the core of the training. Therefore, in the eig week teaching setting, the cognitive exercise focusing on density analysis and s investigation will take two weeks, density/function distribution and detailed desi will take four to five weeks, concentrated drawing performance will take one to tv weeks, and drawing training will run through the whole teaching process.

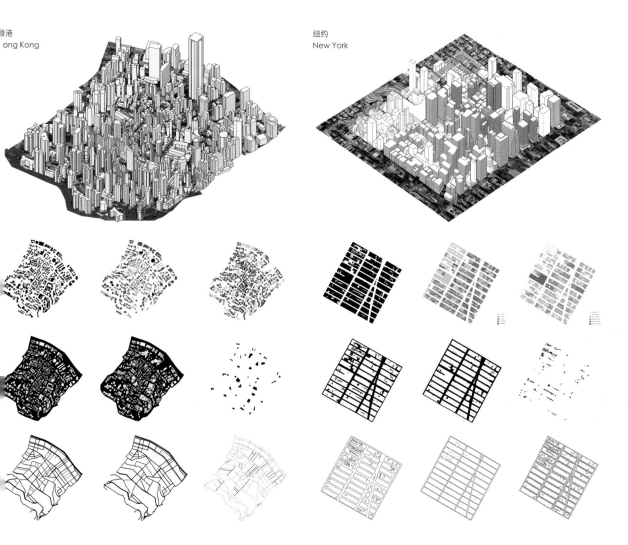

1.密度认知和场地调研

城市空间的质量与交通组织、功能分配、开放空间分布、空间界面等都有紧密联系。"密度分析"的认知环节帮助学生理解"高密度"城市形态的真正含义。通过对东京、纽约、香港、南京等城市高密度片区的形态分析,认知城市切片的街区数量、街道数、建筑数量、建筑高度、建筑面积、功能类型、功能标高等,从而认识到即使在相同的容积率密度之下,交通组织密度、功能分配密度以及空间尺度与形态的差异;同时,通过街景图及空间尺度的分析,学生大致了解到这些指标下人所体验到的城市空间形态。

"场地调研"的认知环节,通过对新街口片区以及设计场地——河西奥体商业片区的实地调研与比较,加强了学生对不同密度下城市空间的真实体验,并发现场地的问题。在分析的同时,学生被要求绘制以城市肌理及外部空间为主要表达对象的空间轮廓平面图、轴测图、分析图表等,通过绘图加深认知。

Density cognition & site investigation

ality of urban space is closely related to transportation organization, function tribution, open space distribution and spatial interface. The cognitive part of "density analysis" helps students understand the real meaning of "high-density" urban form. Through the form analysis of high-density districts in Tokyo, New York, Hong Kong, Nanjing, etc., students can understand the number of blocks, streets, buildings, building height, building area, function type, function elevation, etc. of urban section, so as to realize the difference of density of transportation organization, density of function distribution and spatial size and form even under the same plot ratio density; at the same time, through analysis of street view and spatial scale, students can get a general understanding of the state of urban space experienced by people under these indicators.

In the part of cognition of "site investigation", through the field investigation and comparison of Xinjiekou area and the design site-Hexi Olympic Sports Business District, students' real experience of urban space under different densities is strengthened and the problem of the site is found. While making analysis, students are required to draw the spatial contour plan, axonometric map and analysis chart with the urban texture and external space as the main objects to express, so as to deepen cognition through drawing.

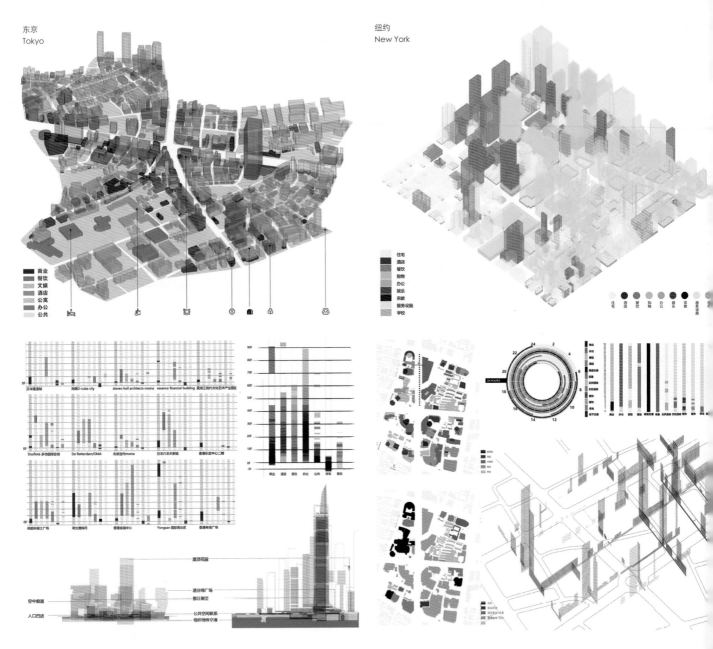

东京
Tokyo

纽约
New York

通过对东京、纽约、香港、南京等城市高密度片区的形态分析，认知城市切片的街区数量、街道条数、建筑数量、建筑高度、建筑面积、功能类型、功能标高等，从而认识到即使在相同的容积率密度之下，交通组织密度、功能分配密度以及空间尺度与形态的差异。

Through the form analysis of high-density districts in Tokyo, New York, Hong Kong, Nanjing, e, students can understand the number of blocks, streets, buildings, building height, building ar function type, function elevation, etc. of urban section, so as to realize the difference of density transportation organization, density of function distribution and spatial size and form even un the same plot ratio density.

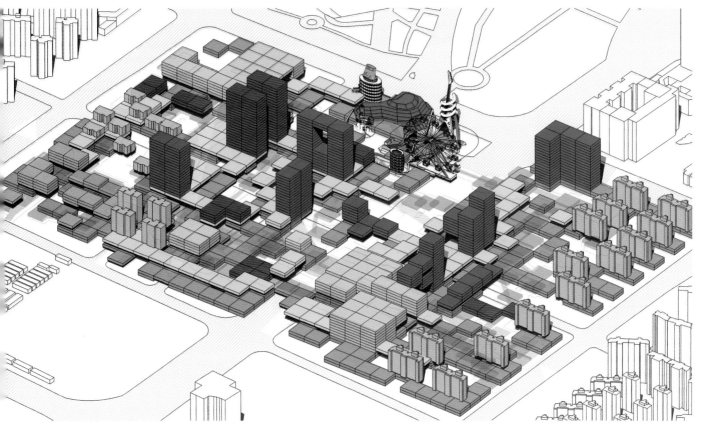

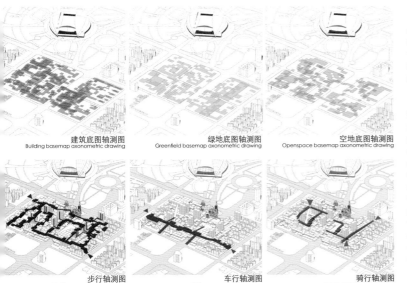

建筑底图图轴测图
Building basemap axonometric drawing

绿地底图轴测图
Greenfield basemap axonometric drawing

空地底图轴测图
Openspace basemap axonometric drawing

步行轴测图
Walking axonometric drawing

车行轴测图
Driving axonometric drawing

骑行轴测图
Riding axonometric drawing

2.密度和功能分配

　　"密度和功能分配"环节训练学生通过城市设计的操作手段与方法——设定场地的路网系统与密度、划分地块、分配用地性质和用地指标、设置公共空间系统等，来获得初步与整体的城市空间形态。每组学生被要求参考前两周所参照的城市切片的密度与功能分配，在建筑总量不能降低的情况下，对设计场地的权属地块及功能进行三维的重组。也就是说，不仅要确定地块及功能的平面位置，也要确定地块及功能的标高位置。随后，确定车行、骑行、步行等交通系统，将这些地块及路网进行串联。

2. Density & function distribution

In "density & function distribution" part, students are trained to obtain the initial and overall urban spatial form through the operation means and methods of urban design—setting the road network system and density of the site, dividing the land, allocating the land use nature and land use index and setting the public space system. Students in each group are required to make reference to the density of urban section and function distribution referred two weeks ago, in the situation that the total number of buildings cannot be reduced, have a three-dimensional reconstruction of the ownership plot and function of the design site. In another word, students should not only confirm the plan position of plot and function, but also confirm the elevation position of plot and function. Later, students should confirm transportation system as driving, riding, walking, etc., and connect these plots and road network in series.

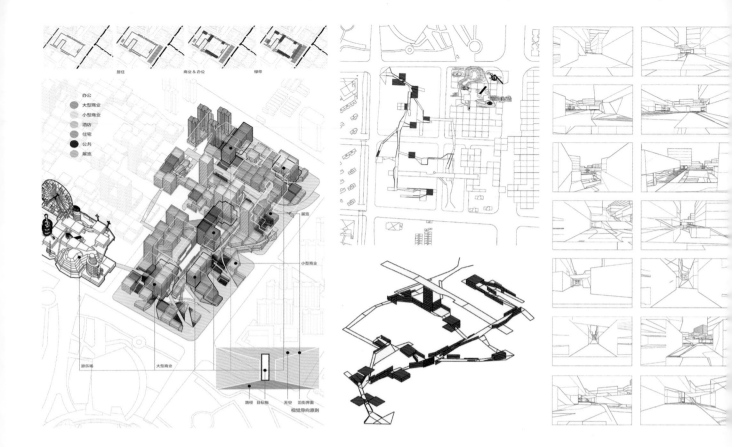

3.性能导向深化设计

实际项目中的城市设计往往需要解决综合问题，作为本科课程训练来说，以明确的单一目标为导向，建立设计手段与目标的有效链接，更易取得好的训练效果。因此，在深化设计环节，课程强调性能导向的优质城市空间的塑造。性能导向包括两个方面：感知性能导向，即获得高质量的、能够具有好的感知体验的城市公共空间；以及环境性能导向，即获得高质量的、具有好的采光性能的城市公共空间。每个学生任选其一作为深化设计的目标，并设定准确的目标意向作为概念，比如"公共路径中能够不断看到有趣的游乐场""公共空间在密集户外活动时间的阳光面最大"等，通过调整空间界面的位置、高度、形式、开口等来实现设计目标。

在感知性能导向组，学生在整体的车行、骑行、步行体系下，以围合感、可视性、可达性、可渗透性等为感知质量的评判标准，以主要公共路径的人眼透视效果为目标，深化设计方案，明确公共空间的具体形式、高度、边界类型，控制构成公共空间界面的建筑边界，并转化为地块的控制导则。在方案深化设计过程中，首先学生被要求选择一条或几条步行公共空间的路径与关键节点，根据设计概念深化设置路径两侧建筑的功能，获得一连串的透视图来体现人在空间中的感知；然后通过对透视图感知质量的评估与优化，调整对透视图效果有影响的建筑高度、边界位置、边界形式，最终获得具有优质感知质量的城市公共空间；最后，将建筑高度、边界位置等转化为地块的控制导则，形成城市设计成果。

3. Performance-oriented detailed design

Urban design in actual projects often needs to solve comprehensive problems. As an undergraduate course training, it is easier to achieve good training results by taking a clear single goal as the orientation and establishing an effective link between design means and goals.

Therefore, in the part of detailed design, the course emphasizes the building of performance-oriented quality urban space. Performance orientation includes two aspects: perceptual performance orientation, namely, obtaining high-quality urban public space with good perceptual experience; and environmental performance orientation, namely, obtaining high-quality urban public space with good lighting performance. Each student selects one as the objective of the detailed design, and sets precise target intention as the concept, such as "interesting amusement park can be constantly seen in public path " "public space has the largest sunlight surface in the dense outdoor activity time", etc., and realizes design objective through adjusting the position, height, form, opening, etc. of spatial interface.

In group of perceptual performance orientation, under the overall driving, riding, walking system, students take sense of enclosure, visibility, accessibility, permeability, etc. as the standard of judging perceptual quality, take visual perspective effect of main public path as the objective, detail the design scheme, define the particular form, height, border type of public space, control the building boundary that constitutes public space interface, and transform into the plot control guide. In the process of detailed design, students are required to choose one or several paths and key notes of walking public space first, set functions of buildings on both sides of the path according to the design concept, obtain a series of perspectives to reflect people's perception in the space; then, by evaluating and optimizing the perceptual quality of perspectives, adjust the building height, boundary position and boundary form that affect the effect of perspectives and finally obtain the urban public space with high perceptual quality; finally, building height, boundary position, etc. are transformed into the plot control guide to form the urban design results.

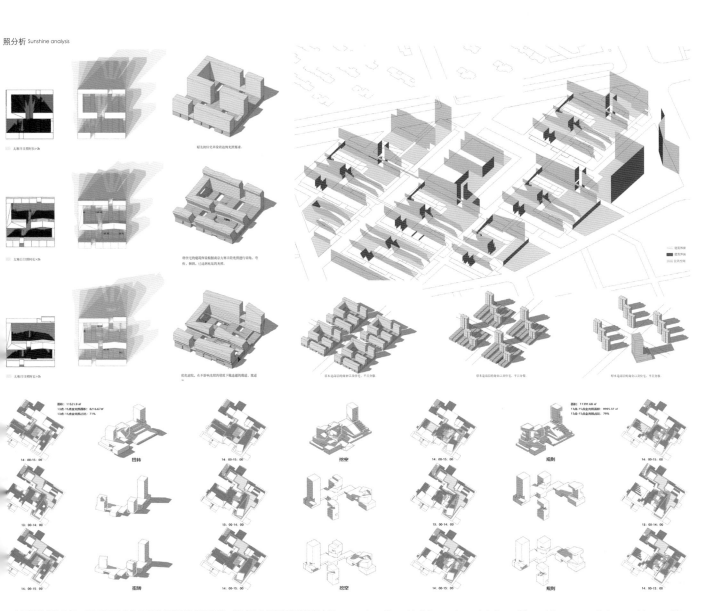

在环境性能导向组，以日照条件作为环境性能质量的评价标准，通过对公共空间的建筑围合界的形式、高度、位置的控制，实现日照条件的优化，并将重要的建筑边界转化为地块控制设计导则。

在方案深化设计过程中，首先要求学生在公共路径上选择三个主要的活动广场作为空间界面的化对象，采用Ecotect软件作为分析工具计算广场区域特定活动时间段（比如15:00—17:00）的照时数现状，并以此为日照条件控制依据进行方案优化。在方案的优化过程中，同学们通过采用不同时间点的日照光线角度的计算以及建筑形体的架空、退让、倾斜、旋转等设计操作，在获得场日照条件改善的同时实现对城市公共空间设计的进一步细化。

he group of environmental performance orientation, students take sunshine condition as the luation standard of environmental performance quality, through the control of the form, height and position of building enclosure interface of the public space, optimize sunshine condition, and transform important building boundary into the plot control design guide. In the process of detailed design, students are required to choose three main activity squares in the public path as the optimization object of space interface firstly, use Ecotect software as the analysis tool to calculate current situation of sunshine hours during specific activity period in the square area (e.g., 15:00-17:00), and optimize scheme on this basis. In the process of optimizing scheme, students further detail urban public space design while improving square sunshine conditions by calculating sunshine ray angle at different times and design of overhead, setback, tilt, rotation, etc. of building form.

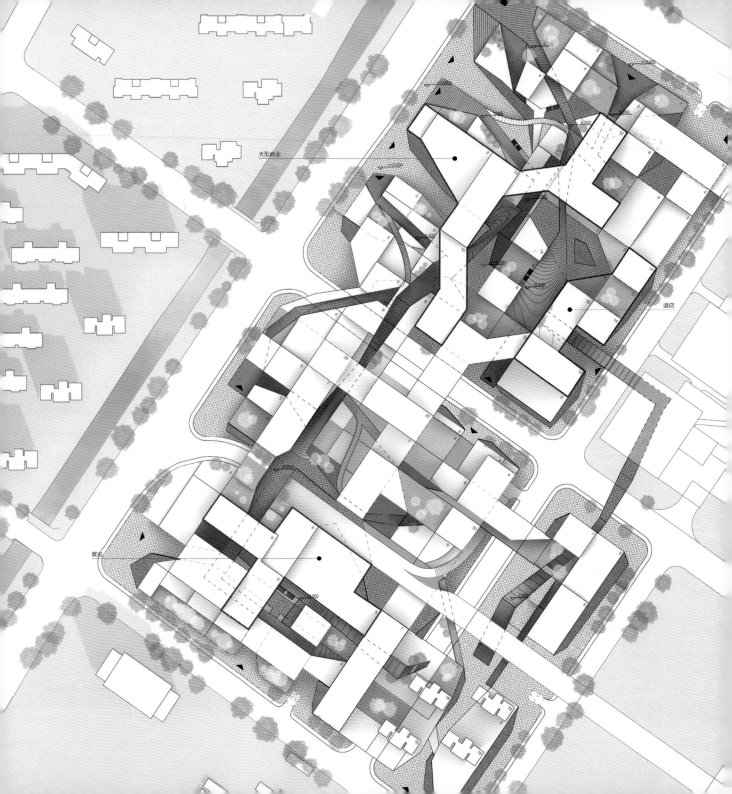

大型商业

酒店

展览

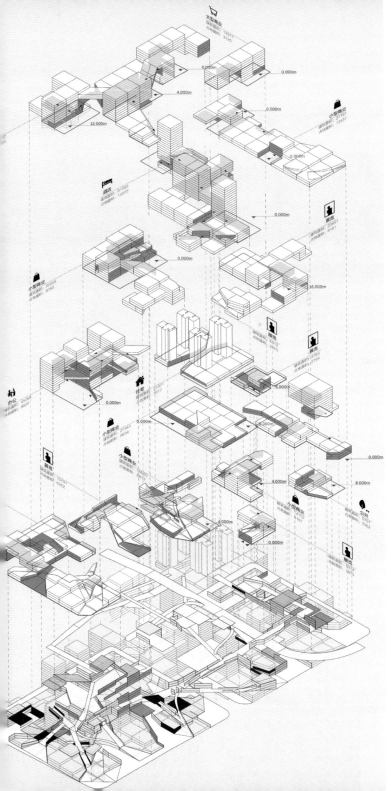

4.图示与表现

伴随城市形态认知、整体设计、深化设计整个过程的，还有城市设计专门的图纸表达训练。城市设计的决策依据是区位、交通、资源及场地的问题及目标，城市设计的操作对象是路网体系、功能分配以及空间界面等，城市设计的目标是获得高质量性能的城市公共空间。因此，城市设计从分析到决策到成果的呈现，都需要针对上述内容进行图纸表述，这对绘图法的训练提出了要求，这项训练贯穿在八周的整个教学过程中，绘图变成了推动认知与设计的工具。

在"密度认知"与"场地调研"环节，训练学生对城市肌理的形态图示与分析能力。除了常规的对城市肌理的轴测图、总平面图的表达，特别设置了分要素的平面图绘制任务，分别是建筑实体黑白图/建筑高度梯度图/建筑基底面积梯度图、外部空间白黑图/公共空间/内院空间、道路/机动车道路/步行道路图等三组内容，还要求了对功能平面特别是高度上的分配的分析图示。通过这些图的绘制，加强学生对肌理形态要素的认知与理解。

在"密度和功能分配"环节，以轴测图为主要图示手段，除了表达整体地块的功能分配与路网设置之外，每个地块单元及建筑类型也被要求用轴测图绘制出来。在"性能导向深化设计"环节，感知性能导向组强调以序列街景图为主要操作对象进行设计的优化，环境性能导向组强调以阳光模拟计算结果图为依据进行设计的优化。通过这些图纸的要求与设定，帮助学生获得深化设计的科学方法。

4. Diagrams and expression

Along with the whole process of urban form cognition, overall design and detailed design, there is special drawing expression training of urban design. The decision-making basis of urban design is the problems and objectives of location, traffic, resources and site. The operation objects of urban design are road network system, function distribution, and spatial interface, etc.. The objective of urban design is to obtain urban public space with high quality performance. Therefore, urban design, from analysis to decision making to the presentation of results, requires drawing expression of aforesaid content, which proposes requirements for the training of drawing method. This training runs through the whole eight-week teaching process, and drawing becomes a tool to promote cognition and design.

In the part of "density cognition" and "site investigation", students' ability in form diagrams and analysis of urban texture is trained. Besides conventional expression of axonometric drawing and general plan of urban texture, plan drawing task with elements is especially provided, respectively black and white drawing of building/building height gradient map/building basement area gradient map, external space white/black drawing of external space/public space/inner courtyard space, road/motor vehicle road/pedestrian road map, etc., analysis diagrams of function plane, especially the height allocation is also required. Through these drawing, students' cognition and understanding of texture form are strengthened.

In the part of "density & function distribution", axonometric drawing is used as the main graphic means. Besides the expression of function distribution and road network setting of the whole plot, each plot unit and building type are also required to be drawn by axonometric drawing. In the part of "performance-oriented detailed design", the group of perceptual performance orientation emphasizes design optimization with the main operating object of sequence street view map, and the group of environmental performance orientation emphasizes design optimization on the basis of sunshine simulation results. Through the requirements and settings of these drawings, students are helped obtain scientific method of detailed design.

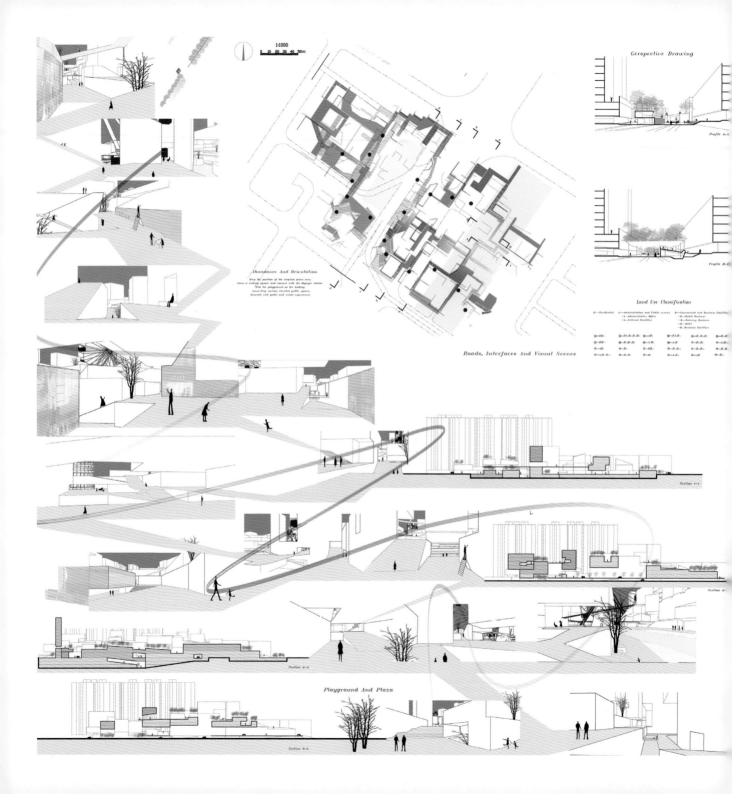

1:1000
0 10 20 30 40 50m

Gerspective Drawing

Profile A-A

Profile B-B

Abundance And Orientation

Keep the position of the original green area,
Form a reading square and connect with the Olympic station.
With the playground as the landing,
connecting various elevated public spaces.
Generate rich paths and visual experiences

Roads, Interfaces And Visual Scenes

Land Use Classification

Section c-c

Section d-d

Section a-a

Playground And Plaza

Section b-b

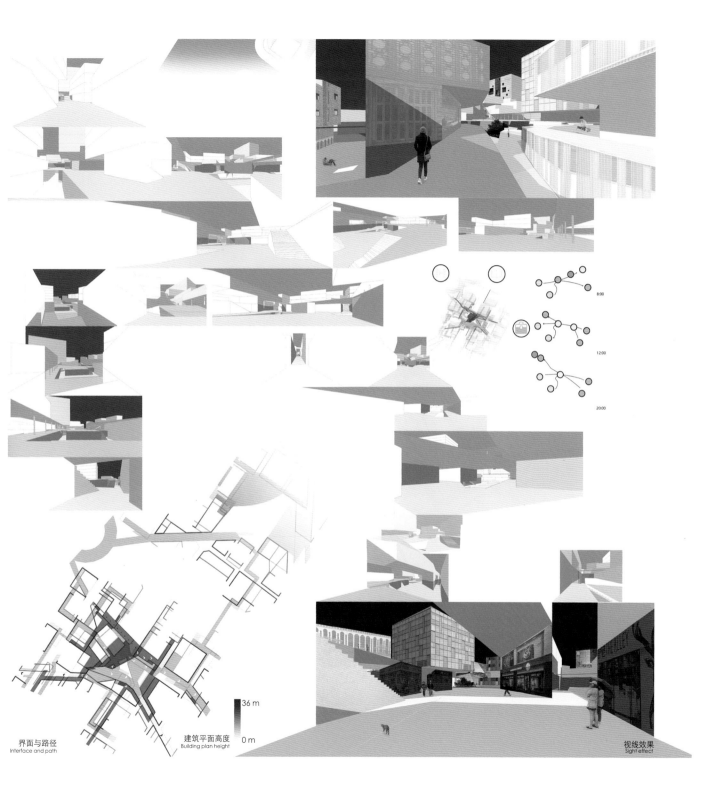

8:00

12:00

20:00

36 m

0 m

界面与路径
Interface and path

建筑平面高度
Building plan height

视线效果
Sight effect

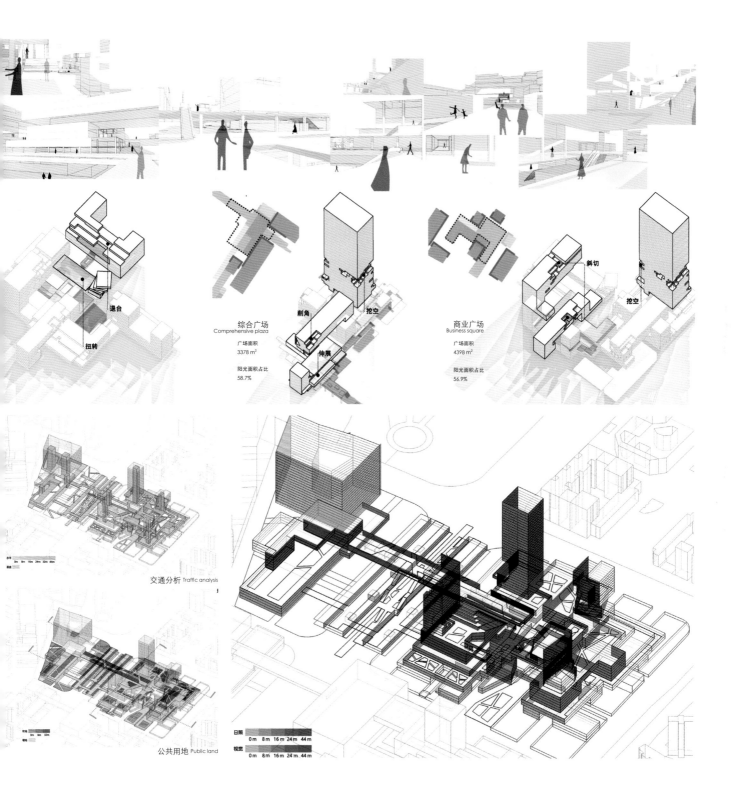

综合广场
Comprehensive plaza

广场面积
3378 m²

阳光面积占比
58.7%

退台

扭转

削角

挖空

伸展

商业广场
Business square

广场面积
4398 m²

阳光面积占比
56.9%

斜切

挖空

交通分析 Traffic analysis

公共用地 Public land

0m 8m 16m 24m 32m 44m

日照
0m 8m 16m 24m 44m

视觉
0m 8m 16m 24m 44m

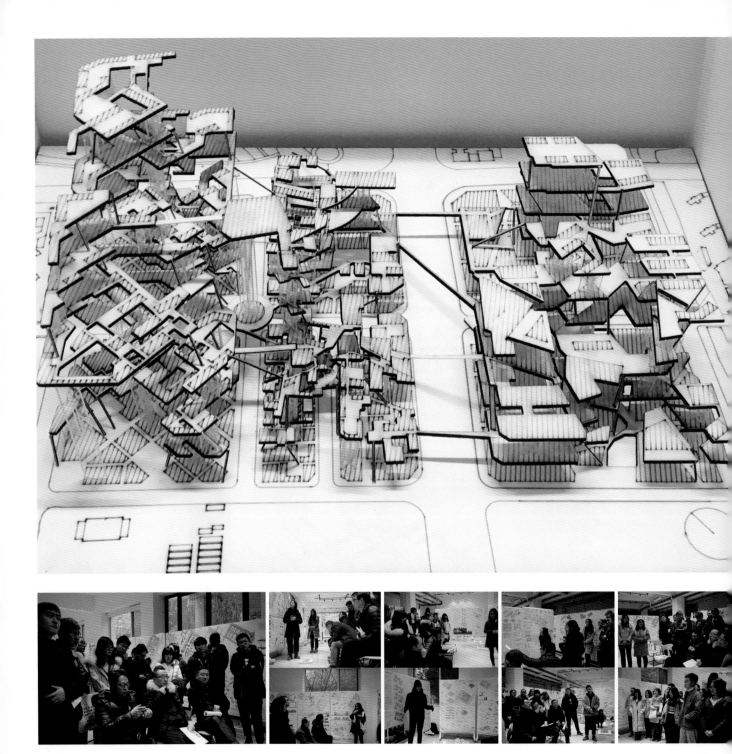

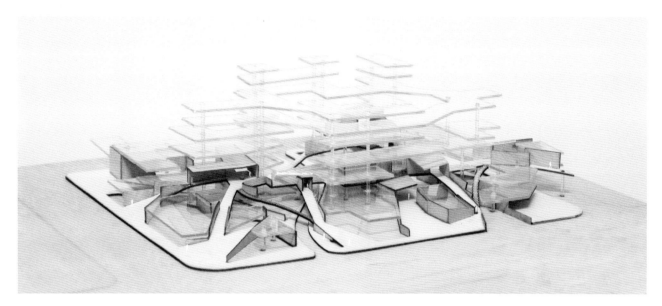

5.小结

    在本科的教学体系中，城市设计是一种专门的设计训练，固然在空间形态设计方法层面延续建筑学的知识，但在认知、决策、成果表达层面与传统的建筑设计又存在显著差异。在此意义上，通过空间形态的创新来提高城市土地的承载能力，不仅是建筑形体——界面空间的创新，更是有依据、有目标的设计手段的合理利用：即在同等的高密度条件下，通过合理地分配功能的位置与高度，再配以有效的交通组织使功能可达。以此为前提，通过建筑界面形体关系的调整，塑造高质量的感知性能、采光性能的城市空间。

5. Summary

In the undergraduate teaching system, urban design is a special design training. Although the knowledge of architecture is continued at the level of spatial form design method, there are significant differences with traditional architectural design at the level of cognition, decision-making and results expression. In this sense, improving carrying capacity of urban land through innovation of space form is not only the innovation of the architectural form - interface space, but also a rational use of well-founded and targeted design means: under equal high density condition, through rational allocation of position and height of functions, coupled with effective transportation organization to make the functions accessible. On this premise, through adjustment of building interface form relationship, urban space with high-quality perceptual performance and sunshine performance can be created.

数字化设计与建造
DIGITAL DESIGN AND CONSTRUCTION
吉国华

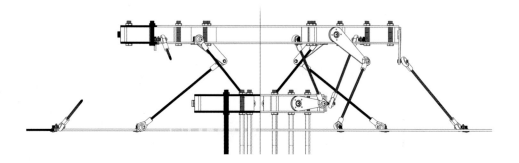

回溯建筑的发展史，给建筑带来根本性前进的是新材料与新技术在建筑上的应用，如工业革命之后玻璃与钢的工业化直接推动了现代建筑的发展。从巴黎美术学院到包豪斯，建筑学的发展是建立在不断地对自我体系的突破之上。基于从标准单元入手构造动态建筑结构的装配实践，本毕业设计涵盖传感器与单片机技术、信号电控制技术、动力输出与机械运动传导、数字建筑交互与表现设计及建造等训练计划，旨在通过训练学习科学有效的数字技术方法，研究数字渗透的建筑设计、面向动态建造的真实问题和数字建筑设计的现实意义。在将本科所学知识融会贯通的基础上，理解设计与当下新的数字技术的关系和研究其对于设计的价值。

1.教学阶段

阶段一：原型分析——借助原型的设计推演

鉴于目前的细分专业系统下的建筑学教育体系没有容纳机械设计与电路控制的相关知识基础，因此由教师所主导的、由成熟机械结构原型出发的案例型教学成为必然途径。首先，这种方法可以迅速地学以致用，快速地将抽象理论与实践相结合。其次，由教师挑选的具有结合潜力的机械原型可以帮助学生快速进入设计推演的角色，运用建筑设计的系统思维借助原型进行设计思考。在此阶段，一方面教师对选取的并联机械原型的工作原理、现有的使用场景以及与建筑学相结合的研究案例做讲解，学生在对原型进行初步了解与资料搜集的基础之上做选择。另一方面，针对单片机编程与控制展开案例教学工作坊，在较短时间内使学生掌握电机控制。学生在初步掌握

电机控制技术之后对实验模型进行测试，并对自己所选取的机械原型进行控制测试从而对所选取原型的动态属性有更加深入的了解。

阶段二：动态的意义——设计反思

传统建筑大部分时间处于相对静止的状态，除了门窗等构件，其余绝大部分在满足使用功能调整时才会做出调整，而且这种被动调整也反映了建筑在绝大多数使用情况下的固有状态。在动态的结构所针对的一些特定使用场景，动态所产生的主动化带来的空间与使用上的新意义是需要进行反思的，互动建筑并不是简单地照搬机械原型，而是通过对机械原型运动重新组合创造新的空间体验、使用场景甚至功能。此阶段，学生需要对所掌握的并联机械原型进行组合设计，反思经过组合之后的机构原型的动态特征是否可以产生有意义的使用场景，这是一个探索的过程，其核心在于设计意义或者说设计价值的产生。针对一种或几种确定的使用场景，学生自定比制作动态展示模型，进入1:1的模型设计与制作过程，对结构动态模型要求能够按照使用场景对运动状态进行切换与调整。

阶段三：创新——再思考与知识的迁移

在完成面向使用场景的设计与展示模型之后，学生被鼓励进一步对动态结构带来的跨学科知识冲击进行再思考，在抽象层面考量互动技术与机械结构是否可以驱更深层次设计体系，使得学生将原本陌生的机械结构知识在理解、掌握并用于设计后，进一步向研究层面迁移。这种知识迁移可以指向原型系统本身，也可以指向建学更大尺度的使用层面，例如城市系统。

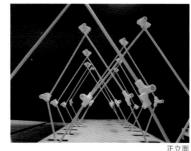

正立面
Facade

案一
Programme 1

方案二
Programme 2

方案三
Programme 3

侧面
Sideview

2.作业实例

(1) 斯图尔特城

罗紫璇的作品"斯图尔特城"将经典的机械结构斯图尔特平台进行变形，利用圆的形式特点将它拓展成层层嵌套的联动结构。这种结构可以在纵向与横向两个方向行叠加，从而产生不同的形式特点。每层圆环被位于上一层圆环的六个转动电机直控制，每一层的运动相互关联。在智能建筑层面的指导意义上，动态结构依照斯图特平台结构纵向叠加形成塔楼。塔楼由内外两层组成，内层提供使用空间，楼板仅上下运动与旋转运动，每层楼板外伸的圆环提供可以自由运动的公共活动空间，产倾斜的坡道、高度渐变的廊道、阳台、雨篷等不同的空间形式。在未来智能城市层的指导意义上，动态结构横向扩展作为一个人造空间站。空间站轨道的移动基于斯尔特平台具有六个自由度的并联运行结构的原理，可以控制平台在一定幅度内进行意位置的改变。每个圆环代表一座城镇的运行轨道，光源位于圆环的中心。城市通不断自转产生的离心力模拟重力，轨道通过倾斜和平移而与光源产生相对位置的改，从而形成昼夜和四季。

(2) 弹性适应系统

卫斌的作品"弹性适应性系统"在斯图尔特平台基础上合并铰链节点，达成可以许空间变形的新型节点，减少自由度，提高位移量，达成可移动的新结构。该新型构几何属性为八面体，通过变形达成移动目的。对杆件和节点进行受力分析可知，缩杆承受沿着杆的压力，而节点则承受来自杆件不同方向的力。这里设计重构遵循拉整体结构原则，即所有力都是沿压杆或者拉索的，这提供了"力始终沿着构件，

且均为拉力或压力"这个思路，将整个几何体的节点和表皮转化成弹性结构和材料，成功地简化了推杆结构，并且确定了弹性节点的连接方式，并重新检查分析了杆件和节点的受力。

(3) 五连杆长屋结构

邱晓宇的作品"五连杆长屋结构"是采用平面五连杆机构对传统长屋剖面结构的动态重构。二度自由平面五连杆并联系顶点的位置坐标与杆件长度和驱动杆的旋转角度有关，通过对同一平面五连杆机构两侧驱动杆进行同步控制，可以实现结构顶点的大幅度偏移。因而通过不同的分组同步控制，长屋结构可以体现不同的空间划分形态。

(4) 悬索张拉整体结构

顾卓琳的作品"悬索张拉整体结构"基于一种质量轻、可变形的六杆二十四索的张拉整体结构，来源于对富勒张拉整体的悬吊结构以及斯内尔森提出的张拉整体编织属性与弓形等效结构的探索。这种结构可以通过施力从类平面转变为类球体。作品"舞苗"通过步进电机带动转盘，牵引张拉结构的杆件，从而驱动整个空间单元产生竖向变化。装置将张拉结构悬挂在三角形钢框架中，结构衔接构件同时作为垂直向的连接装置，用以放置电机和转盘，控制拉线的走向。在三角框架及张拉结构的连接节点的制作中，使用了三维打印技术以实现节点的整体性，电焊增加结构的稳定性。通过感应张拉结构底面的距地距离以及人与装置的位置关系，判断出四种互动情境（靠近、坐下、通过、欣赏）对应步进电机的旋转的速度和方向，以实现装置不同角色间的转变。

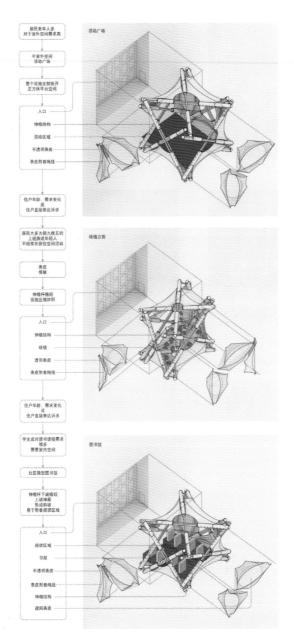

活动广场

居民老年人多
对于室外空间需求高

半室外空间
活动广场

整个设施全部张开
正方体平台空间

入口
伸缩结构
活动区域
半透明表皮
表皮附着网线

住户年龄、需求变化
或
住户直接表达诉求

绿植立面

居民大多为朝九晚五的
上班族或年轻人
不经常在居住空间活动

表皮
植被

伸缩杆缩短
收拢压缩整体体积

入口
伸缩结构
绿植
透明表皮
表皮附着网线

住户年龄、需求变化
或
住户直接表达诉求

图书馆

学生或对读书读接需求
增多
需要室内空间

社区微型图书馆

伸缩杆下端缩短
上端伸展
形成斜坡
易于附着阅读区域

入口
阅读区域
书架
半透明表皮
表皮附着网线
伸缩结构
遮阳表皮

Looking back at the development history of architecture, it is the application of new materials and new technologies in architecture that brought the fundamental advancement to architecture. For example, after the industrial revolution, the industrialization of glass and steel directly promoted the development of modern architecture. From Beaux-arts to Bauhaus, the development of architecture is based on the continuous breakthroughs of the self-system. Based on the assembly practice which constructs dynamic building structures from standard units, this graduation project covers training programs such as sensor and single-chip computer technology, signal electrical control technology, power output and mechanical motion conduction, digital architecture interaction and performance, and design and construction, etc., aiming at studying the digitally permeated architectural designs, real problems for dynamic construction and the practical significance of digital architectural design through the training and learning of scientific and effective digital technology methods. On the basis of a comprehensive study of the subject, this graduation project helps us understand the relationship between design and the current new digital technology, and studies its value to design.

1.Teaching stages

Stage 1: prototype analysis — design deduction with prototype

In view of the fact that the current architecture education system under the subdivided professional system does not contain the relevant knowledge base of mechanical design and circuit control, the case-based teaching approach led by the teacher and based on the mature mechanical structure prototype becomes an inevitable approach. First of all, this method can be quickly applied to combine the abstract theory with practice. Secondly, the mechanical prototypes selected by the teachers with combinative potential can help students quickly enter the role of design deduction, and use the systematic thinking of architectural design to design and think with the help of prototypes. At this stage, on the one hand, the teacher explains the working principles of the selected parallel mechanical prototype, the existing usage scenarios and the research cases combined with architecture orientation. After a preliminary understanding of the prototype and the data collection, the students make a choice. On the other hand, the case teaching workshop for the programming and control of single-chip computer enables the students to master the motor control in a short period of time. After the initial mastery of the motor control technology, the students can test the experimental model and carry out the test of control for the selected mechanical prototype, which gives a deeper understanding of the dynamic properties of the selected prototype.

Stage 2: dynamic meaning — design reflection

The traditional buildings are relatively static for most of the time, except for the doors and windows, the vast majority of the components only make adjustments when they do not meet the functions. This passive adjustment also reflects the inherent state of the building under most conditions of use. In some specific usage scenarios for dynamic structures, the new meaning of space and usage brought about by the proactive changes generated by dynamics needs to be reflected. The interactive architecture is not simply copying the mechanical prototype, but creating new spatial experiences, usage scenarios, and even functions through the recombination of mechanical prototype movements. At this stage, students need to do combination design of the parallel mechanical prototype they have mastered, and reflect whether the dynamic features of the combined prototypes can produce meaningful usage scenarios. This is a process of exploration, and its core is on the generation of design

eaning or design value. For one or several determined usage scenarios, students
reate a dynamic display model with a custom scale, enter the 1:1 model design and
roduction process, and are required to accurately switch and adjust the motion state
ccording to the usage scenarios for the structural dynamic models.

tage 3: innovation — rethinking and knowledge transfer
fter completing the design and display model for the usage scenario, students are
ncouraged to rethink the interdisciplinary knowledge shocks brought about by the
ynamic structure, and whether the interactive technology and mechanical structure
t the abstract level can drive a deeper design system, helps students further transfer
e original unfamiliar mechanical structure knowledge to the research level after
nderstanding, mastery and application for design. This knowledge transfer not only
oints to the prototype system itself, but also to a larger scale of use of architecture,
uch as urban systems.

Job instance
) Stewart city
uo Zixuan's work "Stewart city" transforms the classic mechanical structure Stewart
atform, and uses the form features of the ring to expand it into a layers-nested
nkage structure. This structure can be superimposed in both the longitudinal and
teral directions to produce different form characteristics. Each ring of the layer is
rectly controlled by six rotating motors located in the upper ring, and the motion
each layer is related to each other. In the guiding sense of the intelligent building
vel, the dynamic structure is longitudinally superimposed according to the structure
the Stewart platform to form a tower. The tower is composed of inner layer and
iter layer. The inner layer provides space for use, and the floor is only used for
and down movement and rotation. The ring extended from each floor provides a
iblic space for free movement, resulting in different spatial forms such as sloping
mps, highly gradual corridors, balconies, and canopies. In the guiding sense of the
nart city level in the future, the dynamic structure is laterally extended as an artificial
ace station. The movement of the space station track is based on the principle of
parallel operating structure with six degrees of freedom of the Stewart platform,
hich can control the platform to change at any position within a certain range. Each
ng represents the running track of a city, and the light source is located at the center
the ring. The city simulates gravity by the centrifugal force generated by constant
tation. The track changes its relative position with the light source through tilting
d translation, thus forming a day and night and four seasons.

) Elastic adaptability system
ei Bin's work "Elastic adaptability system" incorporates hinge nodes to achieve
new type of node that can tolerate space deformation on the Stewart platform,
ducing the degree of freedom and increasing the displacement to achieve a new
ovable structure. The geometric property of this new structural is an octahedron,
d the movement is achieved by deformation. Through the force analysis of the rods
d nodes, the telescopic rods are subjected to the pressure along the rods, while the
des are subjected to forces from different directions of the rods. Here, the design
construction follows the principle of tensegrity structure, that is, all the forces are
ong the pressing rod or the cable, which provides the idea that "the force is always
ong the member, and is always tension or pressure". This transforms the nodes
d skin of the entire geometry into elastic structures and materials, successfully
mplifies the push rod structure, determines the connection method of the elastic
des, and re-examines the forces on the members and nodes.

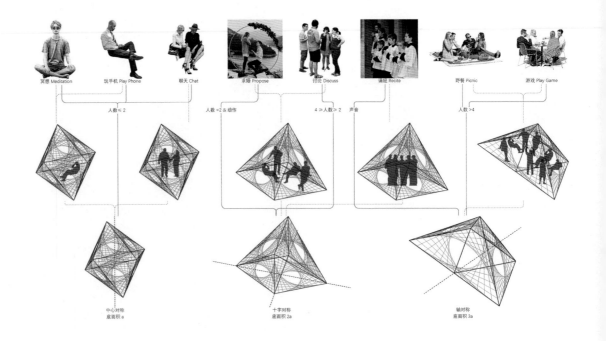

冥想 Meditation　　玩手机 Play Phone　　聊天 Chat　　求婚 Propose　　讨论 Discuss　　诵经 Recite　　野餐 Picnic　　游戏 Play Game

人数 ≤ 2　　　　人数 ≈2 & 动作　　　4 ≥人数 ≥ 2　声音　　　　　人数 >4

中心对称　　　　　　十字对称　　　　　　　　　轴对称
底面积 a　　　　　　底面积 2a　　　　　　　　　底面积 3a

(3) Five-link longhouse structure

Di Xiaoyu's work "Five-link longhouse structure" is a dynamic reconstruction of the traditional longhouse section structure using a planar five-link mechanism. The position coordinates of the apex of the two-degree-of-freedom planar five-link parallel system are related to the length of the rod and the rotation angle of the driving rod. By synchronously controlling the driving rods on both sides of the same planar five-link mechanism, a large shift of the apex of the structure can be achieved. Therefore, through different group synchronized control, the longhouse structure can reflect different spatial division patterns.

(4) Suspension tensegrity structure

Gu Zhuolin's work " Suspension tensegrity structure " is based on a lightweight, deformable six rods and twenty-four cables tensegrity structure, which is derived from the exploration of the suspension structure of Fuller tensegrity structure and

the tensegrity structure weaving property and bow equivalent structure proposed Nelson. This structure can be transformed from a type-plane to a type-spheroid force. The work "Wu Jian" drives the turntable through a stepping motor to pull t rods of the tensegrity structure, thereby driving the entire space unit to produce vertical change. The device suspends the tensegrity structure in a triangular ste frame, and the structure connects the members, while simultaneously serves a vertical connecting device for placing the motor and the turntable to control t direction of the cables. In the fabrication of the connection nodes of the triangu frame and the tensegrity structure, three-dimensional printing technology is used realize the integrity of the nodes, and the electric welding increases the stability the structure. By sensing the distance from the ground of the bottom of the tensegr structure and the positional relationship between the person and the device, seve interactive situations (close, sit, pass, and appreciate) corresponded to the speed a direction of the rotation of the stepping motor are judged, realizing the transitions different roles of the device.

模型问题 & 解决方法
Model problem & solution

所遇问题&解决方案&时间节点

模型照片
Model photo

模型轴测
Model axonometric

撑杆如何实现 (3.7)

节点如何制作 (3.12)

空隙如何围合 (3.11)

灯光如何加入并匹配空间 (4.29)

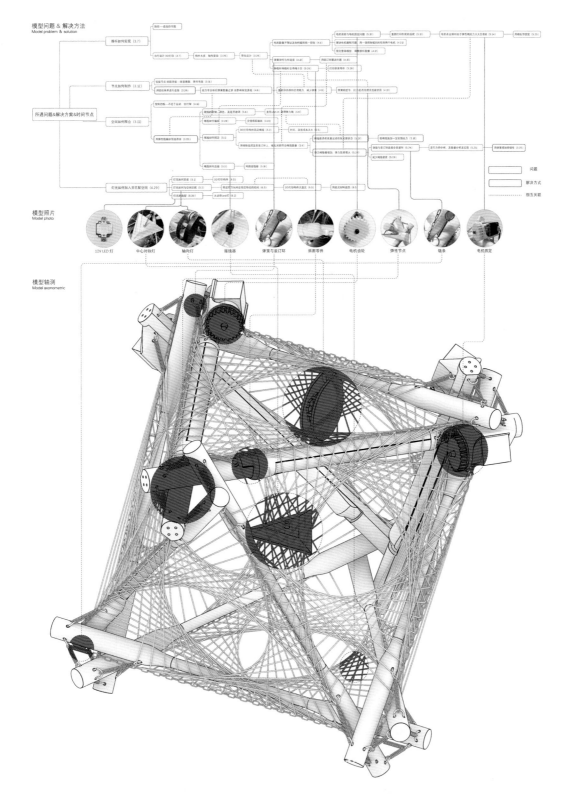

12V LED 灯　　中心对称灯　　轴向灯　　接线器　　弹簧与旋订环　　斜面零件　　电机齿轮　　弹性节点　　链条　　电机固定

问题
解决方式
----- 相互关联

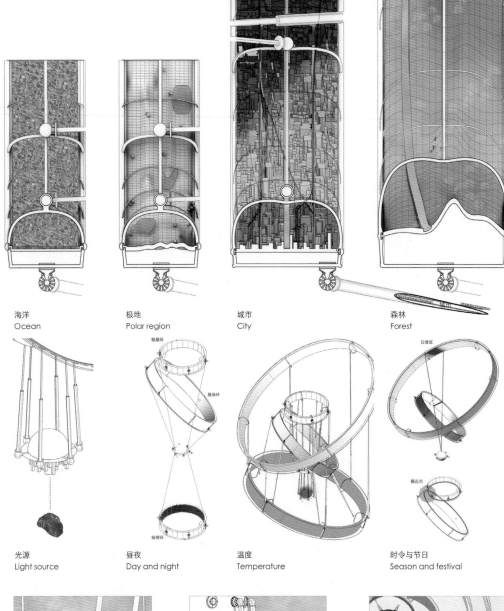

海洋
Ocean

极地
Polar region

城市
City

森林
Forest

光源
Light source

昼夜
Day and night

温度
Temperature

时令与节日
Season and festival

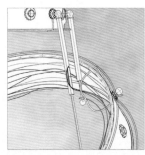

重力
Gravity

交通
Traffic

权力层级
Hierarchy of power

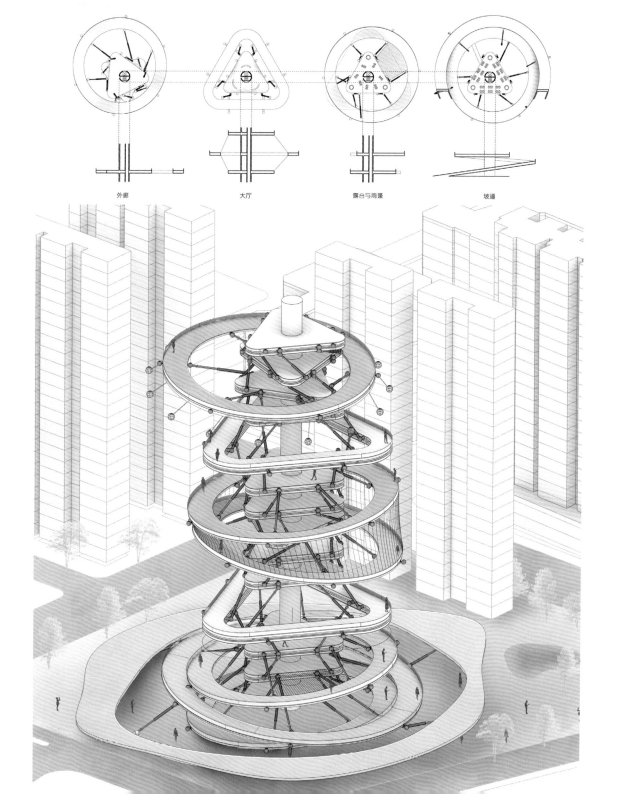

外廊 大厅 露台与雨篷 坡道

77

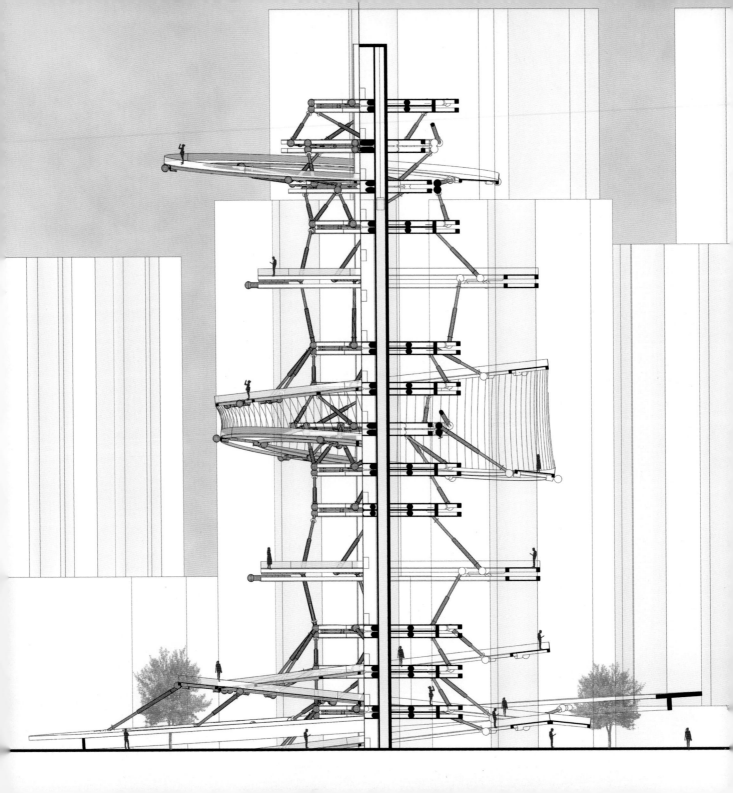

 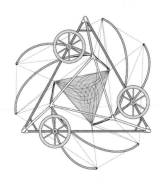 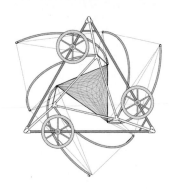 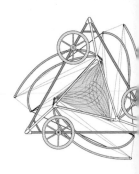

运动过程平面图
Plane graph of the movement progress

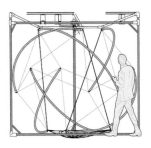  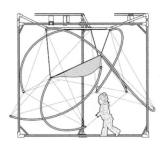 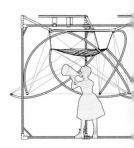

运动过程立面图
Elevation graph of the movement progress

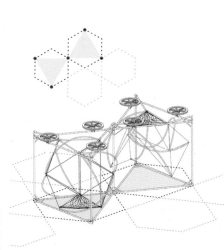 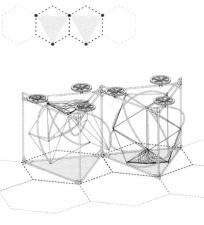 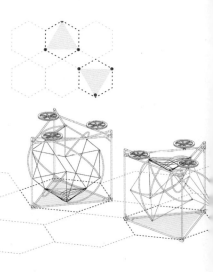

组合方式图解
Diagram of combination

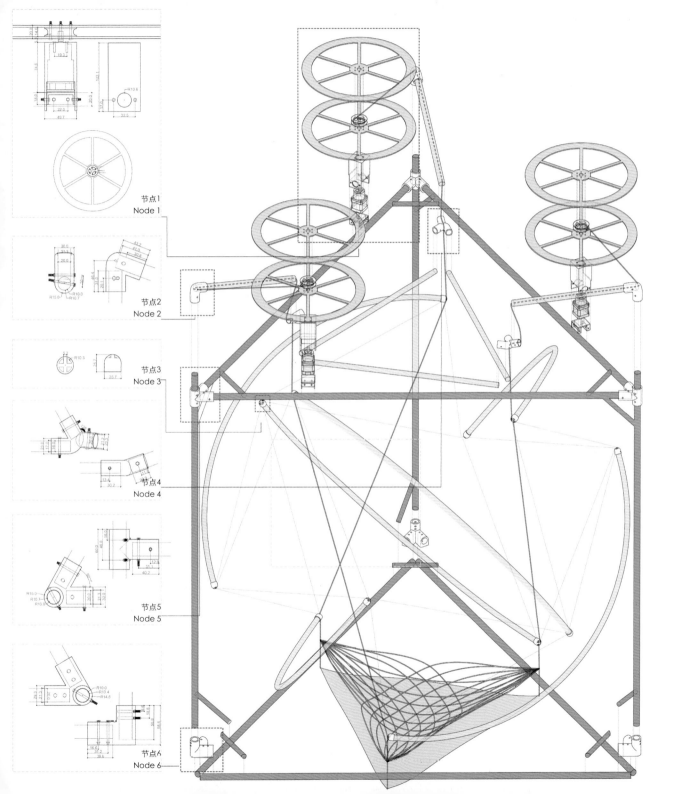

节点1
Node 1

节点2
Node 2

节点3
Node 3

节点4
Node 4

节点5
Node 5

节点6
Node 6

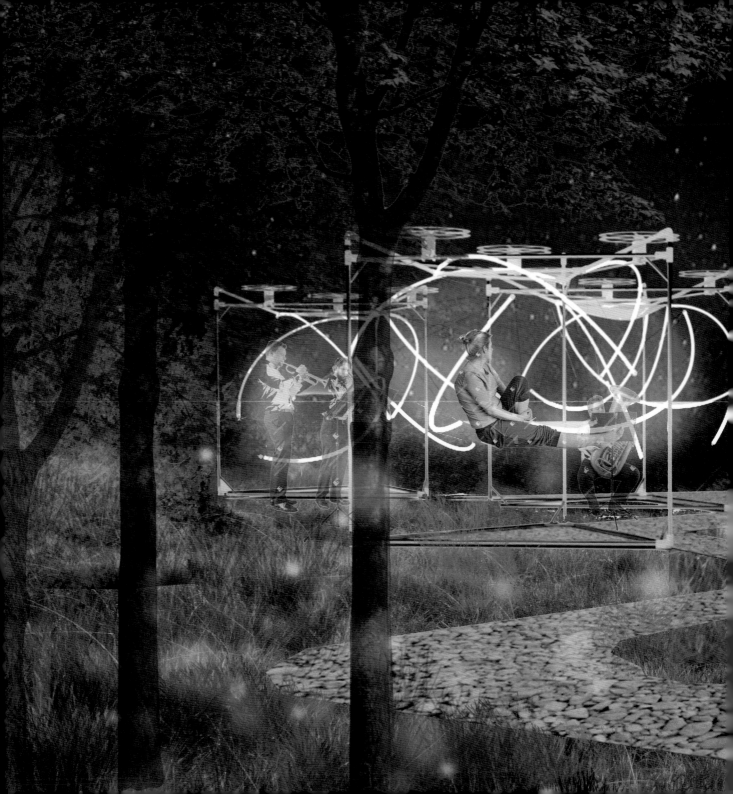

中国木建构文化研究
STUDIES IN CHINESE WOODEN TECTONIC CULTURE
赵辰

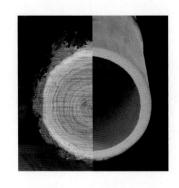

安排
Program

阶段	A. 木构教学+工作坊	B. 工业化竹材工作营设计研究	C. 2018工业化竹材生态建造工作营
Phases	Wooden Tectonic Lectures & Studios	Design Studio with Industrialized Bamboo	Construction Workshop with Ecological Material of Industrialized Bamboo 2018
	5周 5 weeks	6周 6 weeks	4天 4 days

教学 Lectures	讲座一至五 Lecture1~5	评图与讨论 Review and Discussion	"节点设计研究"讲评 Review of Joint Design Study	

| 工作坊 Studios | 作业一：六木同根 Studio1: "Liumu Tonggen" 作业二：木构框架（模型研究）Studio2: Wooden Frame (Model study) | 场地研究 Site Study 方案设计 Design | 施工深化 Construction Design 节点研究 Joint Study | 研讨会开幕 Opening of the Symposium | 建造1 Construction 1 | 建造2 Construction 2 | 竣工典礼 Complet Ceremor |

| 准备工作 Preparations | 模型材料准备 Preparation for Model Material | 工业化竹材下单生产 Production Bill of Industrialized Bamboo | 其他材料下单 Production Bill of Other Material 基础施工 Foundation Construction | 施工准备 Construction Preparations |

第一讲 对全球木建构文化的重新认识

前言: 关于本课方法论的探讨
  • 建构思考作为理论架构与教学程序
  • 设计的理论思维与实际操作的结合

A. 对作为建材的木材之重新评价
  • 费莱切尔和梁思成的认识: "土与木，所有建材之根源"
  • 基于可持续发展理论的理解: 木材作为最生态的建材
  • 木材一直以来就是最方便和最大量使用的材料之一

B. 木构作为世界建筑的传统
  • 欧洲
  • 东亚的深厚传统
  • 美洲的规模化建造
  • 木建构与农业、工业、后工业文明

C. 世界现代木构建筑之新发展
  • 理论与社会的基础
  • 欧洲大师作品所代表的现代木构发展
  • 日本大师的当代木构探索
  • 可持续发展意义的欧洲现代木构案例

Lecture 1   Revaluation for wooden tectonic culture in the world

Forewords: a methodological approach for the course
  • Tectonic thinking as the theoretical framework and teaching program
  • Theoretical thinking combined with practical operation for design

A. Revaluation of the wood as the construction materials
  • Historically, the understanding of Sir. B. Fletcher and Liang Sicheng for earth and wood: "the origin of all construction materials"
  • Today's understanding based upon the theory of "sustainable" development: wood, as one of the most ecological construction material
  • Wood has been one of the most convenient and used materials for human life

B. Wooden construction as a tradition in the World
  • Europe
  • A deep tradition in East Asia
  • Mass construction in America
  • Wooden tectonic with agricultural, industrial, post-industrial civilization

C. New development of modern wooden architecture in the world
  • The theoretical and social base
  • The European masterpieces in modern wooden construction development
  • Japanese masterpieces in contemporary explorations of wooden tectonic
  • Some combination cases of wooden construction development with "sustainability" in Europe

作业一　六木同根
Studio 1　"Liumu Tonggen"

第二讲　中国木建构文化的原则和方法之一：木质材料性与构造
A. 重新审视中国及东亚之木构传统
  • 生态意义的木构传统思考
  • 木材作为自然材料持续性地被人类使用
  • 木材能持续性使用的原因：木材性能、建造传统
B. 木材特性的认知
  • 纤维与单向性
  • 纤维的密度与走向导致木材质的差异
  • 木材的力学特性：单向受力
  • 木材的有效材料特性：线性、杆件
C. 木材的搭接
  • 木材的垒叠搭接：井干
  • 木材的绑扎搭接：绳木
  • 木材的穿插搭接：榫卯
  • 中国榫卯技术的发展
  • 欧洲的榫卯
  • 日本的榫卯
D. 工作室："六木同根"，中国传统木构节点
  • 目的：理解中国传统木构节点的作用
  • 要求：按图制作"六木同根2号"

Lecture 2　The principle and methodology of Chinese wooden tecton
culture: material and construction
A. Rethinking on the tradition of wooden construction in China and East Asia
  • Thinking on wooden construction history with ecological meaning
  • Wood as the natural material sustainably to be used by human being
  • Causes for wood to be sustainably used: features of timber, constructi
    tradition
B. Cognition of the features of timber
  • Fiber and mono-direction
  • Density and directions of fiber deliver the differentiation of timber
  • Physical specification of wood: mono-loading
  • Physical feature of timber: linear, stick
C. The construction of timber
  • Construction with log crossing
  • Construction with tie up
  • Construction with penetration: mortise & tenon
  • Chinese technology development of mortise & tenon
  • European mortise & tenon
  • Japanese mortise & tenon
D. Studio: "Liumu Tonggen", Chinese game of wooden joints
  • Goal: understanding of the Chinese traditional wooden joints
  • Requirements: making "Liumu Tonggen-2" according to the plans

作业二　木构框架
Studio 2　Wooden frame

三讲　中国木建构文化的原则和方法之二：从家具到建筑物

建构意义的中国木构传统之探索
• 中国传统木构体系之结构受力主体
• 木构体系中的官式与民间之异与同

最有效的木结构组织：框架
• 如果砖的结构特点是"拱"，线性杆件木材的结构特点应该是"框架"
• 有效材料的杆件搭接组合成的空间形态
• 杆件之间以榫卯搭接而成的中国的木框架

中国传统木建筑中框架部分与屋面的关系
• 对中国建筑木构体系的古典主义诠释缺少足够的木框架理解
• 穿斗式的木构体系清晰地反映框架意义
• 官式木构体系如何反映框架意义

家具和建筑：木框架作为理性的演变结果
• 木框架的形成：家具的演变
• 木框架的形成：建筑的演变

从家具到建筑：人本的木框架
• 中国建筑的模数：李约瑟的说法
• 勒·柯布西耶的"模度"和中国木建构文化中"间"的意义
• 由人体与框架的关系来确定从家具到建筑的基本原则

工作室-2，从构造到结构单元：木构框架（模型研究）
• 目的：特定的杆件材料（工业化竹材）构成框架的结构单元意义

Lecture 3　The principle and methodology of Chinese wooden tectonic culture: from furniture to building

A. Exploring the tectonic meaning in Chinese wooden construction tradition
   • Main structure of wooden construction system in Chinese tradition
   • Similarity and differentiation between vernacular and official types in wooden construction system

B. A frame, as the most efficient structural organization of timber
   • If bricks want to be an "arch", linear timber sticks want to be a "frame"
   • A special form combined by efficient sticks
   • Chinese wooden frame: constructed by wooden sticks with joints of mortise & tenon

C. The relationship between roof and frame in the Chinese traditional wooden architecture
   • The lack of understanding of wooden frame with classic interpretation of Chinese wooden structure system
   • Column-and-tie system expressed clearly the meaning of wooden frame
   • How to understand the meaning of frame in the official wooden architecture

D. Furniture and building, wooden frame as the rational result
   • Forming the frame: development of the furniture
   • Forming the frame: development of the building

E. From furniture to building, wooden frame unit with humanity
   • Modular system in Chinese building: Joseph Needham's interpretation
   • Le Corbusier's "the Modulor" and the meaning of "Jian" in Chinese wooden tectonics
   • The principle to define from furniture to building with the relationship between frame and human body

F. Studio-2, from construction to structure unit: wooden frame (model study)
   • Goal: defined material (industrialized bamboo) of sticks to be framed as structure unit

工业化竹材（竹集成材）
Industrialized bamboo (laminated bamboo)

第四讲　东亚建造的工业化生态型材料：从木材到竹材
前言：对竹的认知
　　• 竹与人类文明的发展
A. 竹材，作为一种建筑用材
　　• 作为建筑用材之原竹的材料特性
　　• 原竹的材料特性导致的构造特点
B. 可持续发展前提下的竹材开发应用
　　• 与木材相比的竹材开发潜力
　　• 基于生态意义相比木材的使用竹材之优势
C. 原生材料的工业化加工，作为现代化发展的途径
　　• 中国木构传统的现代化发展问题回顾
　　• 木材与木结构的分解与合成
　　• 分解而合成的工业化竹材具有巨大的建材潜力
D. 工业化竹材的建造应用
　　• 建造业的规模化应用意义
　　• 工业化竹材作为结构性材料
结语：工业化竹材结构性能的呈现

Lecture 4　Industrialized ecological building material in Eastern Asian: from wood to bamboo
Forewords: a cognition of bamboo
• Bamboo with the development of human civilization
A. Bamboo, as the construction material
• The physical features of bamboo as construction material
• Construction characters of the physical features of bamboo
B. The application of bamboo material in the situation of sustainable development
• The potential to develop on bamboo compared with wood
• The advantages to utilize bamboo compared with wood based upon ecological meaning
C. Natural material industrialized to be as the way of modernization
• Review to the development of modernization for Chinese wooden construction tradition
• Disassembling and assembling of timber and timber structure
• It has great potentiality in construction materials of industrialized bamboo with assembling after disassembling
D. Application of industrialized bamboo into construction
• Significance of mass applications in construction
• Industrialized bamboo applied as structural material
Conclusion: embody the structure features of industrialized bamboo

2012 南京大学110周年校庆 "中国大学生建造节"
NJU 110 Anniversary , Student Construction Festival, 2012

第五讲　建造实验在教学中的意义

A. 建造与设计
- 古代，建造包含设计
- 现代，设计指导建造

B. 建造与教学
- 师徒制教学中的建造
- 现代主义运动中的建造
- 包豪斯的建造
- 关于建造实验教学的定义
- 当今国际建筑教学中的建造

C. 南京大学的建造实验
- 木构单元体的建造— 2005
- 国际木构工作营 — 2006
- 南京红山动物园建造工作营 — 2007
- 南京大学110周年校庆 "中国大学生建造节" — 2012

2012 南京大学 "中国大学生建造节"
- 场地：南京大学仙林校园
- 功能目的：快递配送站，信息发布等
- 设计策略：贴地，站立，覆盖

D. 南京大学2018建造工作营（工业化竹材）

Lecture 5　Meaning of construction experiment in architectural education

A. Construction and design
- Ancient times, construction involved design
- Modern times, design conducted construction

B. Construction and teaching
- Construction in the craftsmen system
- Construction in the modern movement
- Construction in Bauhaus
- Definition of the construction experiment in education
- Construction in today's architectural teaching

C. Construction experiments in NJU
- Construction of wooden frame unit — 2005
- Cross cultural wooden construction workshop — 2006
- Construction workshop at the Hongshan Zoo, Nanjing — 2007
- NJU 110 Anniversary, Student Construction Festival — 2012

2012 Student Construction Festival, NJU
- Site: Xianlin Campus, NJU
- Objective: express post stop, information deliver, etc.
- Design strategy: landing, standing, covering

D. 2018 Construction Workshop in NJU (industrialized bamboo)

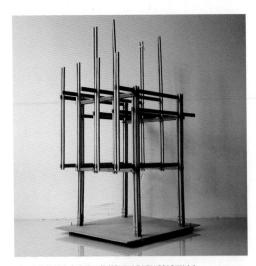

"工业化竹材生态建造工作营"暨AEARU 2018研讨会
"Construction Workshop with Ecological Material of Industrialized Bamboo" for AEARU Symposium 2018

　　"竹塔（工业化竹材多层结构）"，以小断面结构用材，实验高度方向发展的空间多样性。参与者以7周设计整体架构及金属节点；厂家预制构件，并在现场3天内建成。该建造活动成功地实验了现代竹建构的创造性，并激活了校园公共空间。

"Bamboo tower (multi-floored structure with industrialized bamboo)", Nanjing University. The structural elements of light-section were used to construct a tower with space varieties developed in height. The participants designed overall structure including the metal joints in 7 weeks, with prefabrication of all the elements by a factory, and built on the site in 3 days. The workshop successfully experimented potentiality of modern bamboo tectonic culture and activated the public space of the campus.

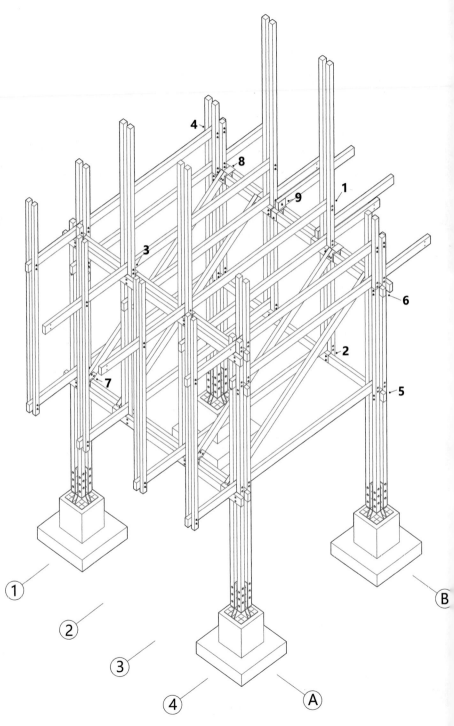

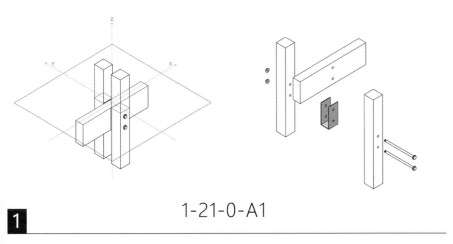

1-21-0-A1

1

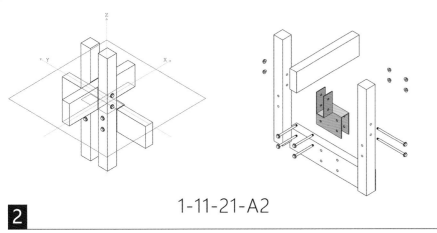

1-11-21-A2

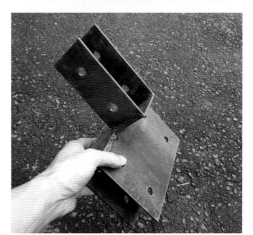

2

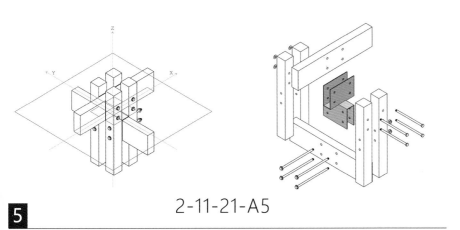

2-11-21-A5

5

MAY 29 / DAY 1
10:00

MAY 29 / DAY 1
15:00

MAY 29 / DAY 1
15:15

MAY 30 / DAY 2
9:40

MAY 30 / DAY 2
9:55

MAY 30 / DAY 2
12:00

MAY 30 / DAY 2
21:45

MAY 31 / DAY 2
0:30

MAY 31 / DAY 2
1:30

MAY 31 / DAY 3
14:00

MAY 31 / DAY 3
15:00

MAY 31 / DAY 3
15:30

MAY 29 / DAY 1
15:45

MAY 29 / DAY 1
19:45

MAY 29 / DAY 1
21:00

MAY 30 / DAY 2
13:15

MAY 30 / DAY 2
15:00

MAY 30 / DAY 2
18:20

MAY 31 / DAY 3
10:45

MAY 31 / DAY 3
11:50

MAY 31 / DAY 3
13:00

MAY 31 / DAY 3
16:00

MAY 31 / DAY 3
17:00

MAY 31 / DAY 3
17:50

竣工典礼（2018年5月31日）
Completion Ceremony（May 31, 2018）

2019国际竹建筑大赛（IBCC 2019）
2019 International Bamboo Construction Competition (IBCC 2019)

　　方案以"竹之器"为设计概念，采用工业化竹材为建筑结构材料的体系化设计与建造：以金属件连接，完全工厂预制加工现场安装。旨在探讨新材料、新技术引领下的未来绿色建筑方向。"器"取"机器"之意，代表了可预制、可复制、快速传播之意。工业化竹材具有满足建筑基本耐久性使用的"器"之能力，是面向未来的生态型的建筑材料，更适于社会大工业生产背景的未来可持续发展要求的建筑事业。

The design concept is based on "bamboo of machine". It used industrialized bamboo as building material for systematic design and construction: the connection is made of metal parts, which are completely prefabricated and installed on site. It is intended to explore the future green building direction under the guidance of new materials and technologies. Machine means prefabricated, copied and spread quickly. Industrialized bamboo has the capability to meet the durability of buildings. It is an ecologic building material for the future, and more suitable for the future sustainable development of social industrial production background.

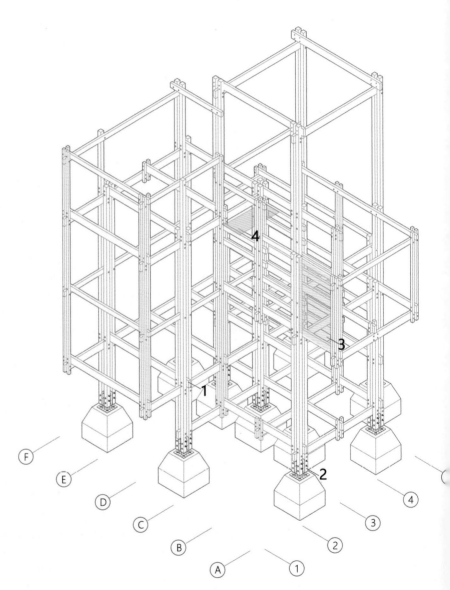

1

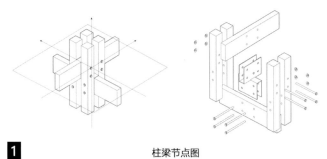

柱梁节点图
Column-beam Joint Detail

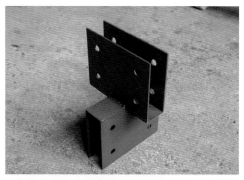

2

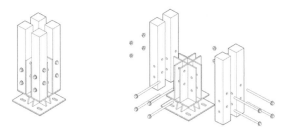

柱脚钢节点图
Foundational Steel Joint Detail

3

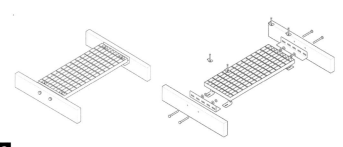

钢格栅板节点图
Steel Grating Plate Joint Detail

4

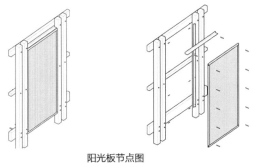

阳光板节点图
Outerbuilding Envelope Joint Detail

JULY 16 / DAY 1
13:30

JULY 16 / DAY 1
15:50

JULY 16 / DAY 1
18:50

JULY 17 / DAY 2
9:20

JULY 17 / DAY 2
10:30

JULY 17 / DAY 2
11:40

JULY 17 / DAY 2
20:00

JULY 17 / DAY 2
21:30

JULY 17 / DAY 2
24:00

JULY 18 / DAY 3
14:20

JULY 18 / DAY 3
15:30

JULY 18 / DAY 3
16:30

JULY 16 / DAY 1
20:30

JULY 16 / DAY 1
20:45

JULY 16 / DAY 1
21:00

JULY 17 / DAY 2
13:40

JULY 17 / DAY 2
15:40

JULY 17 / DAY 2
17:40

JULY 18 / DAY 3
9:40

JULY 18 / DAY 3
11:00

JULY 18 / DAY 3
12:40

JULY 18 / DAY 3
18:00

JULY 18 / DAY 3
19:00

JULY 18 / DAY 3
20:15

竣工典礼（2019年7月18日）
Completion ceremony（July 31, 2019）

垂直城市：一种高密度建筑集群设计研究
VERTICAL CITY: RESEARCH ON DESIGN OF A HIGH DENSITY BUILDING CLUSTER
周凌

课题简述

　　课程研究聚焦中国发展过程中的土地资源紧缺问题以及如何高效利用土地，如何在这样的大背景下提出新的建筑学解决策略。"垂直城市"指一种能将城市要素包括居住、工作、生活、休闲、医疗、教育等装进一个或几个体量里的巨型建筑。在"垂直城市"里，可以提供完备的城市居住、工作和生活功能，如银行、邮局、餐饮、购物、办公等。"垂直城市"拥有庞大的体量、超高的容积率、惊人的高度、少量的占地、高密度的居住人群等特征。"垂直城市"对土地的极度高效利用，可以释放更多空间给城市绿地和居住空间，保证生态绿地的规模化。本课程要求在南京鼓楼滨江下关CBD地块，设计一座垂直城市，主要功能为居住、办公、酒店等，辅助功能自行定义，高度150～200 m。设计既要有很强的超前概念，又要有具体的技术措施和实施策略，具有现实可操作性，包含详细的功能和平面细化。最终提供一个面对现实问题、可以实际实施的建筑方案，完成一次具有高度职业化特征的设计研究训练。

重点概念

　　(1) 垂直城市
　　(2) 垂直社区
　　(3) 垂直村庄
　　(4) 立体花园
　　(5) 空中街道

技术路径：

　　(1) 景观驱动
　　(2) 城市驱动
　　(3) 性能驱动

课程概述

　　"垂直城市——一种高密度建筑集群设计研究"，是南京大学建筑与城市规划学院建筑学专业研究生一年级的设计课程。课程设置考虑两个结合：一是"知识"与"技能"结合；二是"研究性"与"职业性"结合。课程不仅包含技能和职业训练，也有大量自主研究的内容，学生需要经过2～4周的前期研究分析，然后完成一个和研究相关的有研究主题的设计。课程要求既有前沿的高密度城市研究，也考虑具体的工程实践可行性，是一个要求较高的设计课题，也是在现实条件下的"理想化"设计。

1.高密度认知

　　西方城市有几个典型类型，第一种是欧洲19世纪经过规划发展的城市，如巴黎、巴塞罗那，采用街区式建筑，通常6～8层楼，檐口统一高度，高度24 m或28 m，容积率一般在3～5之间。第二种是英美国家普遍采用的，郊区低密度别墅与市中心高密度CBD结合的模式，郊区是独栋或联排别墅，市中心是CBD，CBD密度要求很高，便于人们使用地铁、公交设施到达，减少小汽车的依赖，其中步行距离是关键，所以常见容积率高达5～10之间，甚至更高。第三种是亚洲城市，如东京、香港、新加坡、北京、上海、深圳等，无论城市形态，还是功能，都是一种"斑块状"混杂的模式，大小混杂、高低混杂、商居功能混杂。

　　课程开始时，针对以上主要城市类型进行分析，2～3个同学为一组，研究一些重要城市的规划特点，主要集中在土地利用、交通、城市形态、公共空间几个方面。同学们的分析资料，部分来自国内现有资料，此部分可以建立一个基本的认识；有的小组善于发掘，能够找到最新的国外规划与研究的资料大数据，这就有较大的启发借鉴意义。例如，一个研究小组在对新加坡规划的研究中，看到近年来新加坡最新一轮规划的各种重要举措，包括生活与工作的有效组织、通勤时间的决定性影响等方面，都对我们认识未来城市有较大作用。

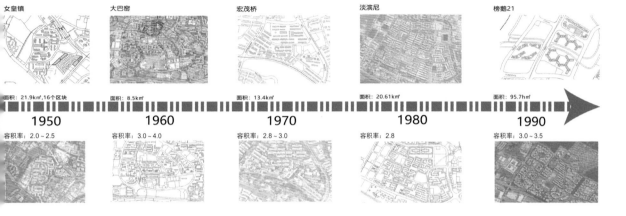

女皇镇
面积: 21.9km², 16个区块
1950
容积率: 2.0~2.5

大巴窑
面积: 8.5km²
1960
容积率: 3.0~4.0

宏茂桥
面积: 13.4km²
1970
容积率: 2.8~3.0

淡滨尼
面积: 20.61km²
1980
容积率: 2.8

榜鹅21
面积: 95.7hm²
1990
容积率: 3.0~3.5

2.中国问题认知

中国城市的土地利用效率相对比较低，近年来，板式住宅对城市产生影响。50 m 度的板式住宅，约18层，容积率2~2.5；100 m高度的板式住宅，约30层，容积率 约3~3.5。其间差别主要是南北维度不同，日照间距要求不同。在历史城市，传统中 国城市的建筑层数较低，以1~2层为主，但是因为采用高密度的方式，容积率不低， 可以达到4层板式住宅的容积率。在一般城市的老城区，还要面对很多20世纪七八十年 代建造的6层板楼。这些板楼效率很低，使城市中心的土地利用效率降低很多。住宅板 楼的大量使用，是中国城市市域面积过大的一个重要原因。因此，紧凑型的建筑组团 和集群变得很重要。课程在讲解了一些基本认知之后，同学们的分析具有一定的针对 性，聚焦更加清晰准确，提高了研究的效率。

3.解决的办法：高密度集群

通过加大建筑容积率，减少城市蔓延，较少私家汽车的使用，增加公共交通的可 能性，提高通勤效率。加大容积率的具体办法有几种，一是提高建筑层数，二是降低 日照要求，三是加大住宅标准层面积，四是使用混合功能。同学们提出了不同的解决 方案，尽量把城市中心住宅高度提高到150~200 m，压缩间距，与办公混合地块使 用。一组同学提出青年公寓与住宅叠加体积，形成立体的高密度社区。另一组提出利 用视线优化原理，错落布局，以得最大化地获得江面景观。利用城市新区广场过大， 使用效率低，在广场中加入青年公寓，提高城市公共空间的土地利用效率。

最后的设计图纸部分，同学们设计构思新颖，各具特色，研究也具有超前性。多 位同学在多轮充分讨论后形成了相对成熟的方案，图纸完成的深度达到了一定的专业 水平。研究生的设计课程与本科不同，其研究性和专业性、深度和职业化程度，都超 过本科设计课程。

Subject introduction

The research focuses on the shortage of land resources in the development process of China and how to use land efficiently, how to propose new solution strategy for architecture in this context. "Vertical city" refers to a huge building which can combine the urban elements including dwelling, work, life, leisure, medical care, education, etc. into one or several volumes. In "vertical city", complete urban dwelling, working and living functions can be provided, such as bank, post office, catering, shopping and office. "Vertical city" has the characteristics of huge volume, super high plot ratio, amazing height, small amount of land cover, high density of residents, etc. The extremely efficient use of land in "vertical city" can release more space for urban green space and living space, and ensure the scale of ecological green space. This course requires the design of a vertical city in CBD plot of Binjiang Xiaguan in Gulou District, Nanjing. The main functions are residential, office, hotel, etc.. Auxiliary functions are self-defined and the height is 150~200 meters. The design should not only have a strong advanced concept, but also have specific technical measures and implementation strategies, to be practical and operable, including detailed functions and plane refinement. Finally, it should provide an architectural scheme that can be implemented in face of practical problems, and complete a design research training with highly professional characteristics.

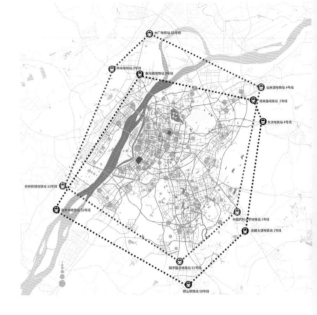

Key concepts

(1) Vertical city
(2) Vertical community
(3) Vertical village
(4) Sky garden
(5) Sky street

Technical paths

(1) Concept from landscape
(2) Concept from urbanism
(3) Concept from performance

Course summary

"Vertical city — research on design of a kind of high-density building cluster" is the first-year design course of the architecture graduate student of the School of Architecture and Urban Planning, Nanjing University. The curriculum design considers two combinations: one is the combination of "knowledge" and "skill"; the other is the combination of "research" and "occupation". The course is not only about skill and vocational training, but also includes a lot of independent research contents. Students need to go through 2~4 weeks of preliminary research and analysis, and then complete a design related to the research with research subject. The course requires not only cutting-edge high-density urban research, but also the feasibility of specific engineering practice. It is a design subject with high requirements, and also an "ideal" design under realistic conditions.

1. High density cognition

There are several typical types of western cities. The first is the planned development of European cities in the 19th century, such as Paris and Barcelona, which adopt block type buildings, usually with 6~8 floors, uniform cornice height, height of 2 meters or 28 meters, and the plot ratio generally between 3~5. The second is adopted widely in Anglo-American countries, the combination of low density suburban villa and high density central CBD mode, suburban single-family or townhouse, CBD in downtown. CBD requires very high density, easy for people to reach by subway bus, so as to reduce the dependence on cars, in which walking distance is important so common plot ratio is as high as 5~10, and even higher. The third is Asian cities such as Tokyo, Hong Kong, Singapore, Beijing, Shanghai, Shenzhen, etc., which are all "patchy" mixed in terms of urban form and function. They are mixed in size, height commercial and residential functions.

At the beginning of the course, based on the analysis of the above major urban types 2~3 students work as a group to study the planning characteristics of some importan cities, mainly focusing on land use, transportation, urban form and public space. A for students' analysis, part of the data is from existing domestic information, and basic understanding can be established in this part; some group is good at diggin data and can find the latest big data of foreign planning and research, which is o great inspiration and reference significance. For example, in the study of Singapor planning, a research team found various important measures in the latest plannin round of Singapore in recent years, the effective organization of life and work, th decisive impact of commuting time and other aspects all played a significant role our understanding of the future city.

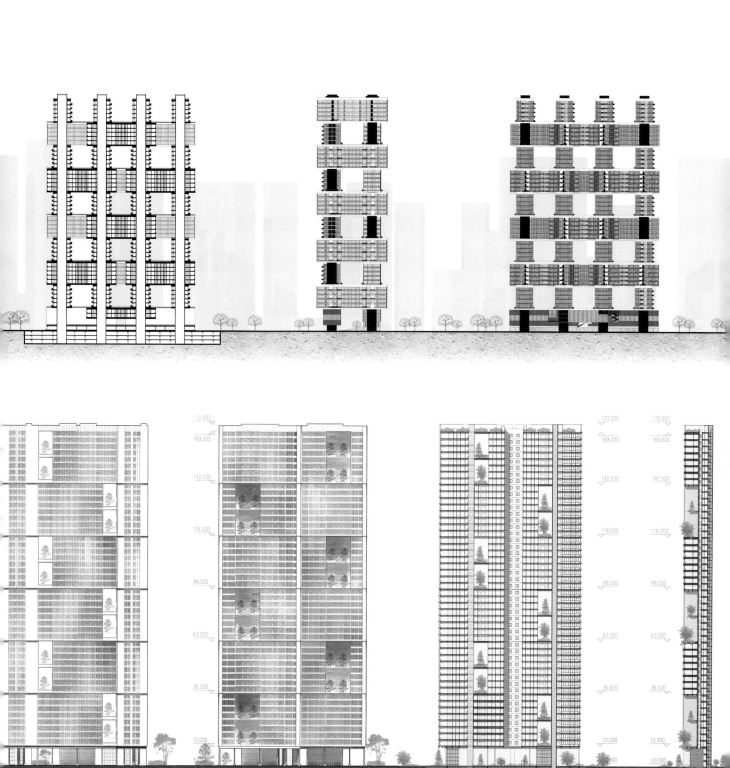

2. Chinese problem cognition

The land use efficiency is relatively low in Chinese cities. In recent years, the slab apartment has an impact on the city. The slab apartment is 50 meters high, with about 18 floors, and plot ratio of 2~2.5; the slab apartment is 100 meters high, with about 30 floors, and plot ratio of 3~3.5. The difference between them is mainly the dimension of the north and the south, sunshine spacing requirements are different. In historical cities, the number of building floors in traditional Chinese cities is relatively low, mainly 1~2 floors. However, due to the high density, the plot ratio is not low, which can reach the plot ratio of 4-floor slab apartment. In the old urban area of city, there are many six-storey slab buildings built in the 1970s and 1980s. The efficiency of these buildings is very low, which reduces the efficiency of land use in the city center. The large use of residential slab apartments is an important reason for the large urban area in China.Therefore, compact architectural groups and clusters become important. After the some basic knowledge was explained in the course, students' analysis was targeted, focused more clearly and accurately, and improved the efficiency of research.

3. Solution: high density cluster

By increasing the building plot ratio, reduce urban sprawl, reduce the use of private cars, increase the accessibility of public transport, and improve commuting efficiency. There are several measures to increase the plot ratio: to increase the number of building floors, to reduce the sunshine requirements, to increase the residential standard floor area, to use with mixed functions. Students proposed different solutions, trying to increase the height of resident in the urban center to 150~200 meters, compressed the spacing, and used with the office mixed land. A group of students proposed to overlap the volume of youth apartment and residence to form a three-dimensional high-density community. Another group proposed to utilize the principle of sight optimization, scatter layout, so as to maximize the river surface landscape. They took advantage of the large square in the new urban area with low use efficiency, added youth apartments in the square to improve the land use efficiency of urban public space.

In the final design drawing part, students' design ideas are novel, with their own characteristics, and their research is also advanced. Most students form a relatively mature scheme after several rounds of full discussion, and the depth of drawing completion reaches a certain professional level. Postgraduate design course is different from undergraduate course, and it is more research-oriented, professional, in-depth and professional than undergraduate design course.

植物
种植土壤
聚酯无纺布
陶粒排水层
C20细石砼
卷材层
保温层
找平层
找坡层
现浇钢砼结构层

石材地面砖
灰浆层
找平层
分割层
撞击隔声层
混凝土楼板
抹灰层

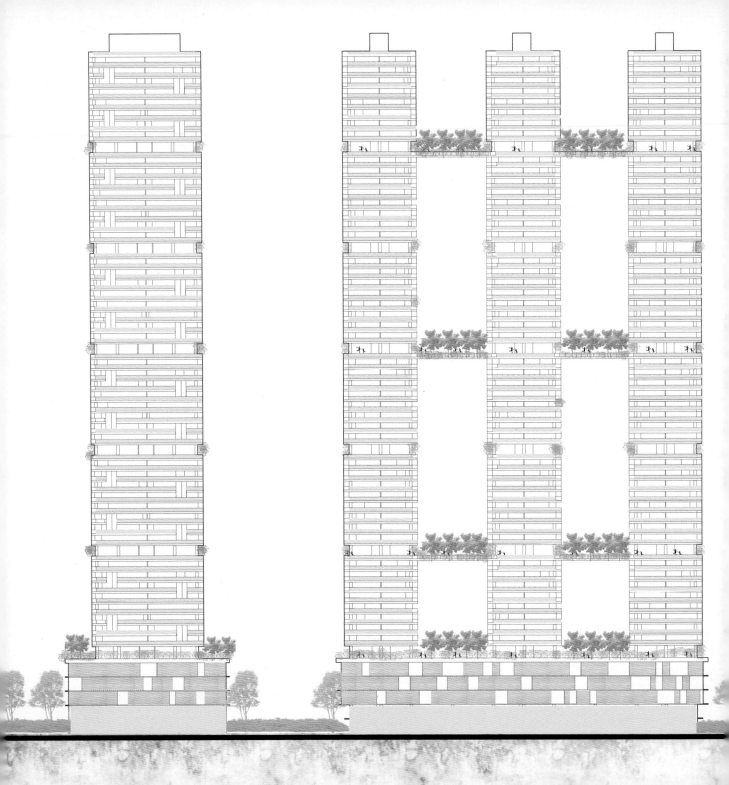

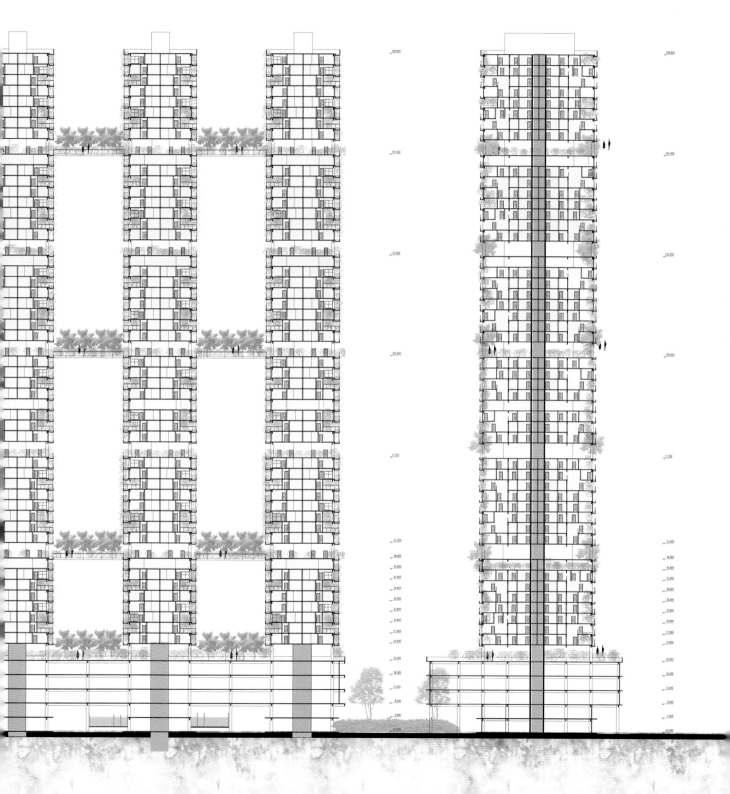

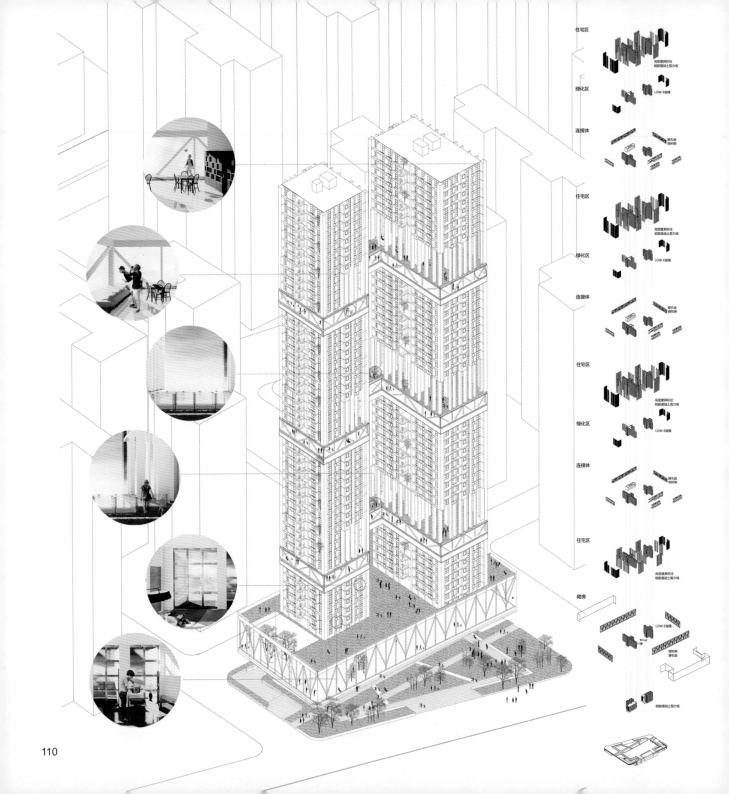

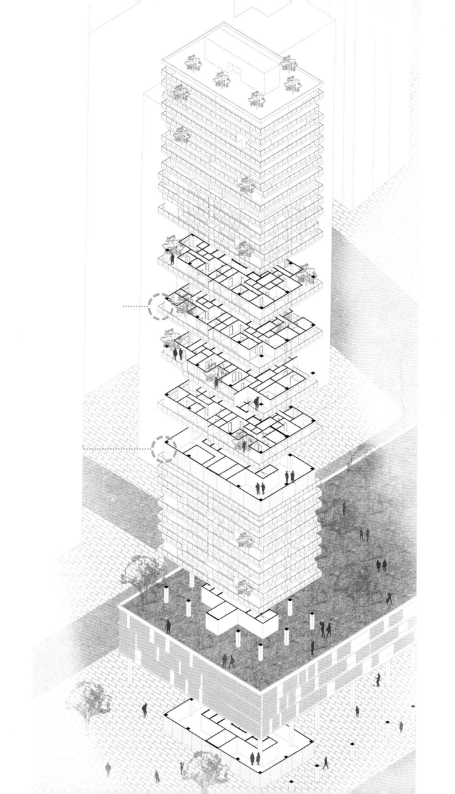

户型A   共60户
125.45㎡
含入户花园
三室两厅,各房间均有江景。

户型B   共60户
154.28㎡
含入户花园
多房房户型,主要房间西向江景。

户型C   共140户
65.8-70.3㎡
根据阳台组合共有2种户型
一室一厅

户型D   共140户
54-55.72㎡
根据阳台组合共有2种户型
一室一厅

户型E   共520户
57.98-64.86㎡
根据阳台组合共有4种户型
一室一厅

户型F   共240户
77.9㎡
景观、日照条件相对较差
集约的三室两厅户型

江景住宅单元

青年公寓单元

公租住房单元

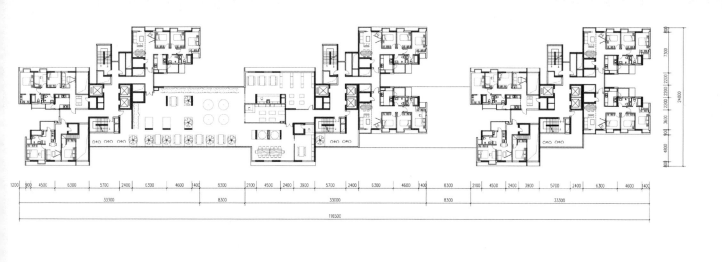

| 1200 | 900 | 4500 | 6300 | 5700 | 2400 | 6300 | 4600 | 1400 | 8300 | 2100 | 4500 | 2400 | 3900 | 5700 | 2400 | 6300 | 4600 | 1400 | 8300 | 2100 | 4500 | 2400 | 3900 | 5700 | 2400 | 6300 | 4600 | 1400 |

| 33300 | 8300 | 33000 | 8300 | 33300 |

116500

| 1200 | 900 | 4500 | 6300 | 5700 | 2400 | 6300 | 4600 | 1400 | 8300 | 2100 | 4500 | 2400 | 3900 | 5700 | 2400 | 6300 | 4600 | 1400 | 8300 | 2100 | 4500 | 2400 | 3900 | 5700 | 2400 | 6300 | 4600 | 1400 |

| 33300 | 8300 | 33000 | 8300 | 33300 |

116500

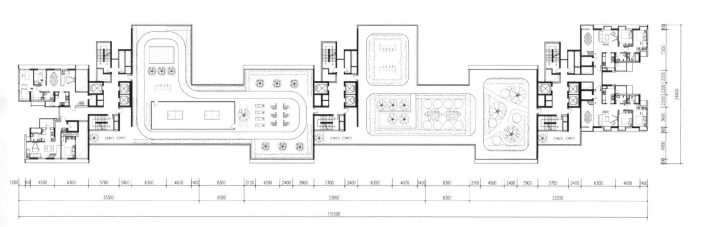

| 1200 | 900 | 4500 | 6300 | 5700 | 2400 | 6300 | 4600 | 1400 | 8300 | 2100 | 4500 | 2400 | 3900 | 5700 | 2400 | 6300 | 4600 | 1400 | 8300 | 2100 | 4500 | 2400 | 3900 | 5700 | 2400 | 6300 | 4600 | 1400 |

| 33300 | 8300 | 33000 | 8300 | 33300 |

116500

城市再生：内边缘转型计划
URBAN REGENERATION: TRANSFORMING INNER EDGES
胡友培

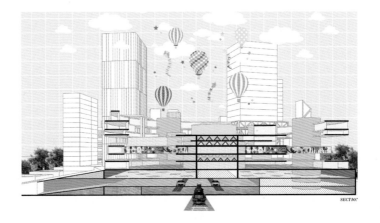

什么是内边缘?

边缘是躁动不安的，在混乱的表面下，是城市蠢蠢的脉动与欲望的张望。

临时性是它的根本属性，地理、物质、人口的变迁，是它永恒的主题。

边缘是开放的，欢迎一切可能，是探险者、违规者与野草的乐园。

边缘也是模糊的，城市、乡野的二元属性，在这里随机地杂交，产生出一种模棱两可的城市景观。城市设计的专业词汇，如场所、尺度、秩序、界面等，在这里是失语的。

内边缘，是一种独特的边缘景观与类型。它天生携带者边缘的基因，继承了边缘几乎所有品性，却因为某种阴差阳错，被包裹、攀养于城市内部。它的粗俗、廉价、丑陋，与周遭光鲜亮丽的城市相比，是如此的不协调，甚至触目惊心。与此同时，它的野蛮、开放、生机盎然，却又让墨守成规、温文尔雅的城市显得如此保守无趣，甚至平庸。

它是城市的另类与遗忘之地。

它是被囚禁的放逐者。

背景与问题

内边缘的形成——内边缘，本是城市的外缘。在中国城市近几十年跳板式的急速扩张中，由于种种原因（基础设施、土地归属、行政管辖、地理自然等）被逐渐包围进城市内部。当摊开城市扩张的宏伟地图时，才惊讶地发现，它已经位于城市腹地。边缘从此成了内边缘。内边缘的存在合理吗？它会一直存在吗？

内边缘的身份——内边缘的身份是模糊的。一方面具有典型的边缘景观，另一方面又盘踞于城市内部。地理和心理地图的错位，造成认知的困惑，更加剧了其身份界定的难度。在城市用地日渐紧张的当下，应该将它作何利用，为它建构何种新的身份？

内边缘的形式——具有边缘色彩的物质要素（大型基础设施、表浅化城市景观以

及低劣的建筑）在很大程度上依然主导着今天内边缘的空间形式。

与此同时，四周日益高企的地价正对其虎视眈眈。 内边缘作为一个独特的城市物种，面临着随时被同化于周遭的城市机体，而消散于无形的生存压力。

在这里除了续写城市的陈词滥调，还可能赋予它什么样的空间形式及形式的意义？

从曾经的边缘，到当下认知盲区，再迈向前途叵测的未来，内边缘转向何方?

为此，项目需要一种现实主义的关怀，需要一种批判性的立场，更需要一种乌托邦的想象力。

向模棱两可的内边缘都市景观，投向诗意的目光。

在最卑微的城市角落，开出独特的花朵。

What is the urban inner edge?

The edge is restless. It is the pulsation and desire of the city under the chaot surface.

Temporality is the fundamental attribute of the edge. Its eternal theme consists geography, matter and the change of population.

The edge is open that welcomes everything possible. It is a paradise for explorer violators and weeds.

The edge is also vague. The binary attribute of the city and the countryside randomly hybridized here to create an ambiguous urban landscape. The profession vocabularies of urban design, such as place, scale, order, interface, etc., are aphas here. The edge means no where, non place.

The inner edge is a unique edge landscape and type. It is born with the gene of th

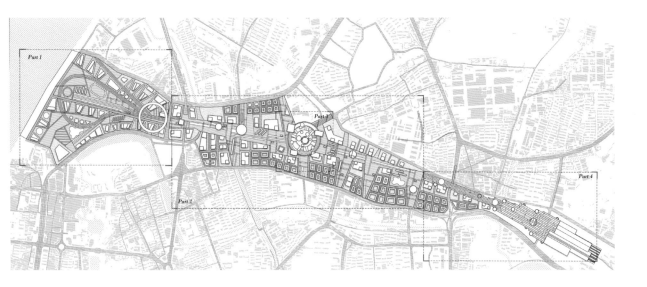

edge and inherits almost all the characteristics of the edge. However, it is wrapped and supported in the city because of some accidental errors. Its vulgarity, cheapness and ugliness are so uncoordinated and even shocking when compared to the glamorous cities around. At the same time, however, its wildness, openness, vigor and vitality also make the conformist, urbane cities so conservative and boring, even mediocre.

The inner edge is the unusual and forgettable place in the city, and it is an outcast that has been imprisoned.

Backgrounds and issues

The formation of inner edge——The inner edge was the outer edge of the city. However, during the springboard-like rapid expansion of Chinese cities in recent decades, the inner edge had been gradually encircled into the inner part of the city due to various reasons (such as infrastructure, land ownership, administrative jurisdiction, geography and nature, etc.). When the grand map of the expansion was spread out, it was surprisingly found that the inner edge was already in the interior of the city. The edge has since become the inner edge. Is the existence of the inner edge reasonable? Will it always exist?

The identity of inner edge——The identity of inner edge is vogue. On the one hand, it has a typical edge landscape. On the other hand, it is also confined to the interior of the city. The geographical and psychological misplacement has caused cognitive confusion and increased the difficulty of defining the identity of inner edge. At the time when urban land is becoming increasingly tense, how can we use the inner edge? What new identity shall we create for the inner edge?

The form of inner edge——The material elements with edged-color (large infrastructure, superficial urban landscape and shoddy construction) still dominate the space form of the inner edge today to a large extent.

At the same time, the increasingly high price of land is eyeing the inner edge. As a unique urban species, the inner edge faces the survival pressure that may be assimilated into the surrounding urban body at any moment and dissipate into nothing.

In addition to continuing to write the cliches of the city, what kind of space form can we give to the inner edge? What is the meaning of the form of inner edge?

From the former edge, to the current cognitive blind spot, then to the unknown future, where is the inner edge going?

For this purpose, transforming the inner edge needs a realistic care, a critical stance and a Utopian imagination.

Fix the sights to the ambitious urban landscape of inner edge.

Make unique flowers bloom in the humblest city corner.

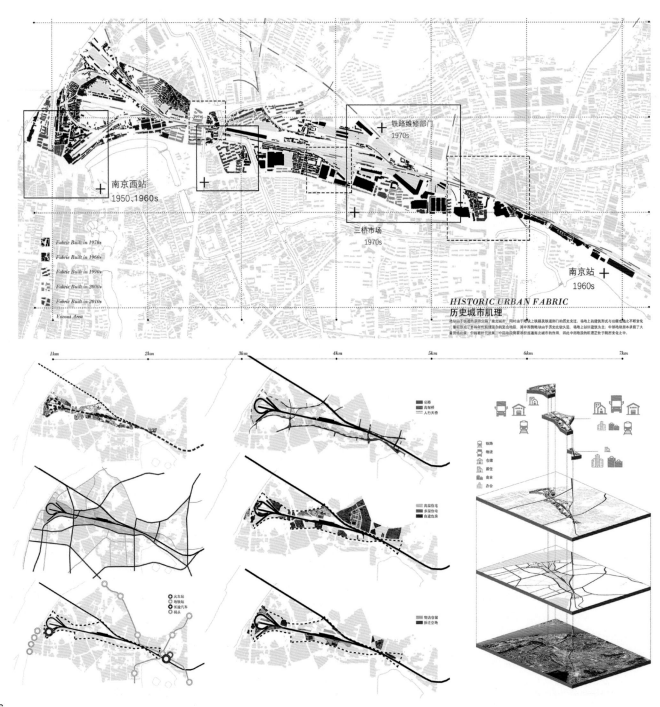

铁路维修部门
1970s

南京西站
1950、1960s

三桥市场
1970s

南京站
1960s

Fabric Built in 1970s
Fabric Built in 1960s
Fabric Built in 1990s
Fabric Built in 2000s
Fabric Built in 2010s

Vacant Area

HISTORIC URBAN FABRIC
历史城市肌理

1km   2km   3km   4km   5km   6km   7km

公路
高架桥
人行天桥

高层住宅
多层住宅
自建危房

火车站
地铁站
长途汽车
码头

物流仓储
拆迁空地

铁路
物流
仓储
居住
商业
办公

116

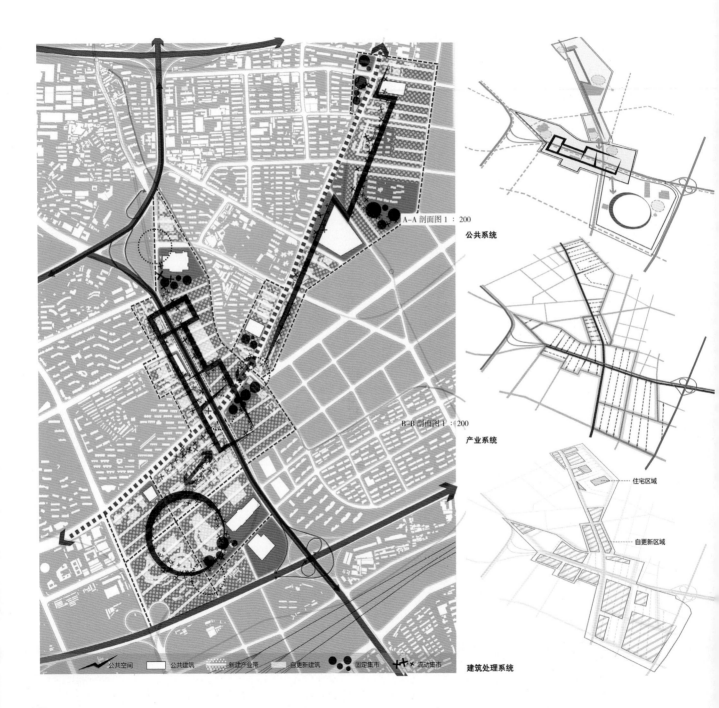

A-A 剖面图 1 : 200

公共系统

B-B 剖面图 1 : 200

产业系统

住宅区域

自更新区域

建筑处理系统

公共空间   公共建筑   新建产业带   自更新建筑   固定集市   流动集市

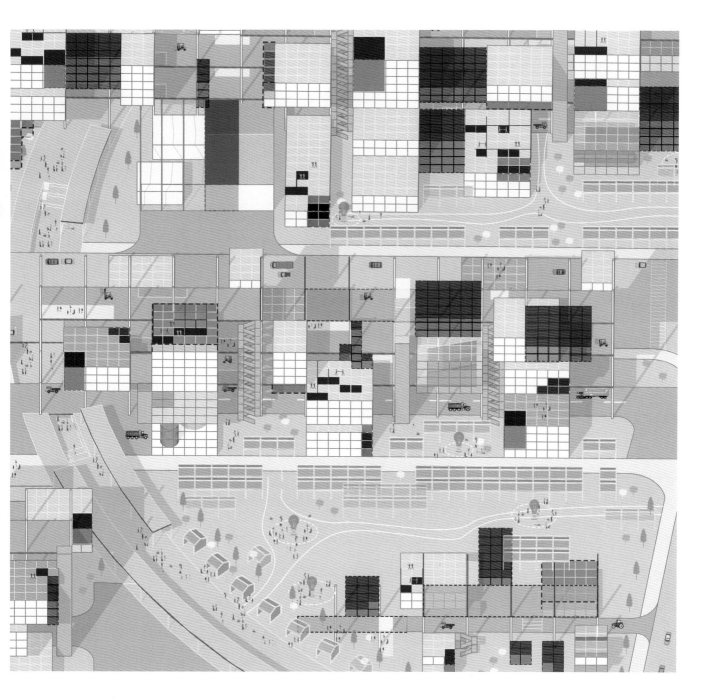

121

supermarket

Basketball court

Kindergarten

Green space

Parking lot

Factory

Vegetable market

Fitness equipment

Electric bike shed

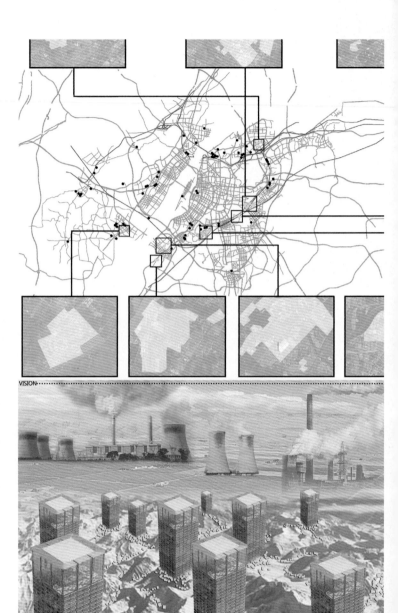

VISION·

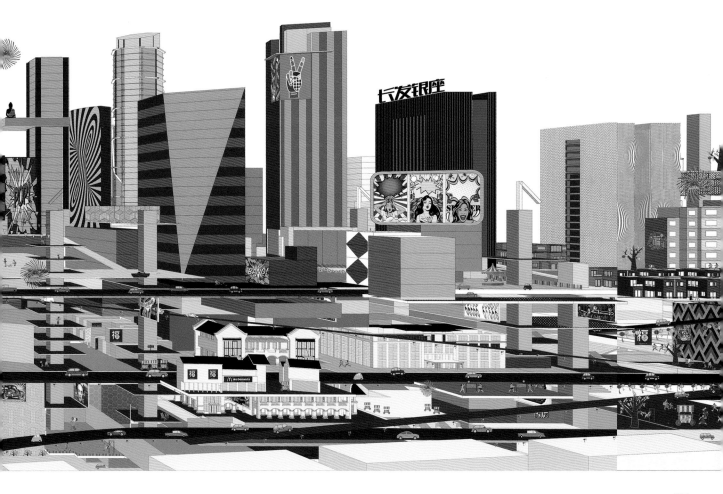

研究生国际教学交流计划
THE INTERNATIONAL POSTGRADUATE TEACHING PROGRAM

南京大学建筑与城市规划学院一直致力于最前沿的学术理论研究，并以之解决实际问题。在研究成果输出的基础之上，学院积极参与国际学术交流并致力于为未来培养建筑和城市规划的高水平人才。为了拓展学院的国际学术交流网络并且建立长期的合作交流体系，2016年南京大学建筑与城市规划学院启动了"研究生国际教学交流计划"（IPTP）。IPTP是一个灵活的邀请教学计划，每年都会有10位专家教授接受邀请前来教学访问。

本年度已经完成的课程有：

1. 参数化图解静力学
   科朗坦·菲韦 孟宪川

2. 建筑技术设计：作为城市再生催化剂的预制功能单元
   罗伯特·博洛尼亚 华晓宁

3. 适应性节点：使用3D打印技术的建筑生态节点设计
   朴大权 钟华颖

4. 应对老龄化的混杂空间：建筑尺度
   米格尔·真蒂尔 傅筱

5. 应对老龄化的混杂空间：城市尺度
   若泽·路易斯·瓦莱乔 傅筱

6. 绿色建筑技术
   拉卡多·布科莱利 郜志

7. 城市形态学：一种城市景观的分析方法
   谷凯 胡友培

Since its foundation, the School of Architecture and Urban Planning, Nanjing University is committed to cutting-edge academic theory courses, in order to address contemporary issues. Based upon research outputs, the school is actively engaged in international academic exchanges and targets at nurturing high-level professionals in architecture and urban planning for the future. In order to further extend its international academic exchange network and to establish a long-term cooperative and exchange mechanism, the school has launched the International Postgraduate Teaching Program (IPTP). The IPTP is a flexible guest-teaching program that includes 10 visiting positions annually.

Programs finished this year include:

1. Parametric Graphic Statics Workshop
   Corentin FIVET, MENG Xianchuan

2. Architectural Technology Design: Prefabricated Functional Unit as Urban Regeneration Catalyst
   Roberto BOLOGNA, HUA Xiaoning

3. Adaptive Joints: Constructing Habitable Structures Using 3D Printed Joinery
   Daekwon PARK, ZHONG Huaying

4. Hybrid Spaces of Negotiation for an Active Ageing: Middle-small Scale
   Miguel GENTIL, FU Xiao

5. Hybrid Spaces of Negotiation for an Active Ageing: Big-middle Scale
   Jose Luis VALLEJO, FU Xiao

6. Green Building Technology
   Riccardo BUCCOLIERI, GAO Zhi

7. Urban Morphology: An Analytical Approach to the Urban Landscape
   GU Kai, HU Youpei

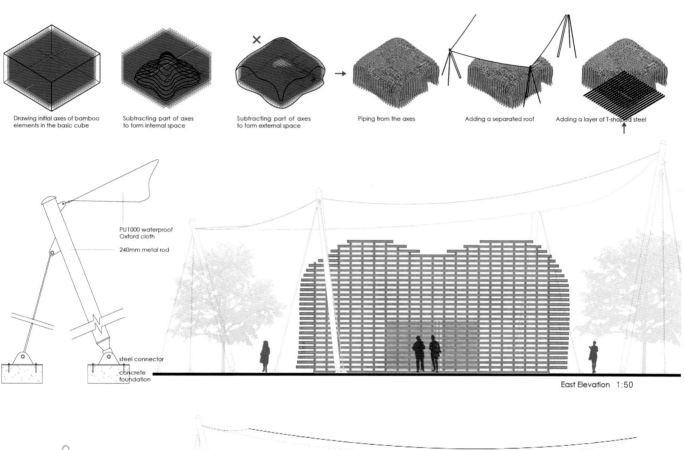

Drawing initial axes of bamboo elements in the basic cube

Subtracting part of axes to form internal space

Subtracting part of axes to form external space

Piping from the axes

Adding a separated roof

Adding a layer of T-shaped steel

PU1000 waterproof Oxford cloth

240mm metal rod

steel connector

concrete foundation

East Elevation 1:50

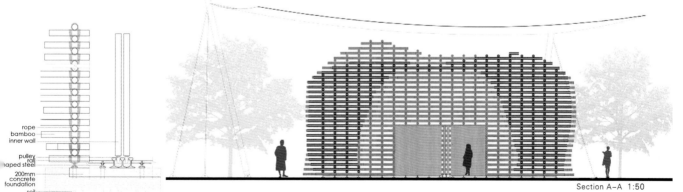

rope
bamboo
inner wall
pulley
rail
T-shaped steel
200mm concrete foundation
soil

Details 1:20

Section A-A 1:50

参数化图解静力学
PARAMETRIC GRAPHIC STATICS WORKSHOP
科朗坦·菲韦　孟宪川

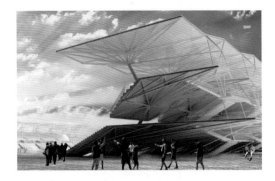

参数化图解静力学设计工作营致力于在设计初期促进建筑与结构的融合。自18世纪建筑学与结构工程专业分工以来，两个学科各自发展迅速，同时也加深了彼此间的隔阂，因此两学科相互融合的议题被不断探索。鉴于建筑师的图形思维方式与结构师的数理思维方式迥异，导致双方难以有效地合作。图形作为媒介的研究方法成为架构两个专业间桥梁的工具。参数化图解静力学正是其中一种前沿的设计方法。

参数化图解静力学设计工作营为建筑师提供了创新的方法与工具，使建筑师们通过数字化方法，操作形式的内部力流和设计结构的几何形式。工作营首先通过案例介绍基本知识。几何技巧加快了形式的建构，以此来广泛地探索静力平衡状态下的结构形态。这些技巧通过参数化的工具（Rhino 3D和Grasshopper）被应用于设计，有利于在设计早期阶段通过客观的标准对结构进行优化。历史上优秀案例的设计过程也将被逐步地揭示。

每个小组的设计成果由四个部分组成：
1.根据设计要求提出初始的概念意向
2.通过不断整合各种设计要素，形成迭代的设计过程
3.利用参数化图解静力学的工具对结构进行优化
4.表达最终的设计结果
到工作营结束，参与的同学具备一定能力：
1.重构设计策略以探索符合低碳要求的结构体系
2.更好地判断给定结构类型的设计/几何的自由度
3.在静态平衡的前提下，针对概念结构体系定制参数化的模型
4.更好地判断如何修改结构的几何形态以增强其结构性能

This workshop on parametric graphic statics attempts to promote the integration between architecture and structure during the early conceptual phase of the design project. Starting from the 18th century the education of architects and structural engineers divided in two separate directions and high-speed developed separately, the gap between each other was deepened, therefore the issue of integration of two professions has been continuously explored. In view of the different between the graphic thinking model of architects and the mathematical thinking model of engineers, it is difficult for both professions to cooperate effectively. Graphic approaches are potentially very good media to bridge such gap between architects and engineers. Parametric graphic statics is one of such frontier design approaches. This workshop presents innovative methods and tools that give the architects the opportunity to control the design of a structural geometry together with its internal flow of forces. The workshop will start with the basics, introducing the general rules by means of practical examples. Geometric shortcuts are emphasized in order to speed up the graphical construction, and hence the wide exploration of structural arrangements in static equilibrium. Implementations of these methods into parametric software tool (Rhino 3D and Grasshopper) are addressed alongside with objective criteria that can be used early in the process to optimize the structure. Historical examples of interactive design processes will also be showcased.

The final presentation of each group includes four parts:
1.Conceptual intension based on the design conditions
2.Iterations of the design process following the integration of multiple design factors
3.Structural optimization with parametric graphic static tools
4.Renderings of the final design
At the end of the workshop, participants are able to:
1.Reproduce conceptual design strategies to explore structural systems that have low-carbon impact
2.Better determine the degrees of design/geometric freedom of a given structural typology
3.Build tailored parametric explorations of conceptual structural systems in static equilibrium
4.Better determine how to modify the geometry of a structure in order to enhance static behavior

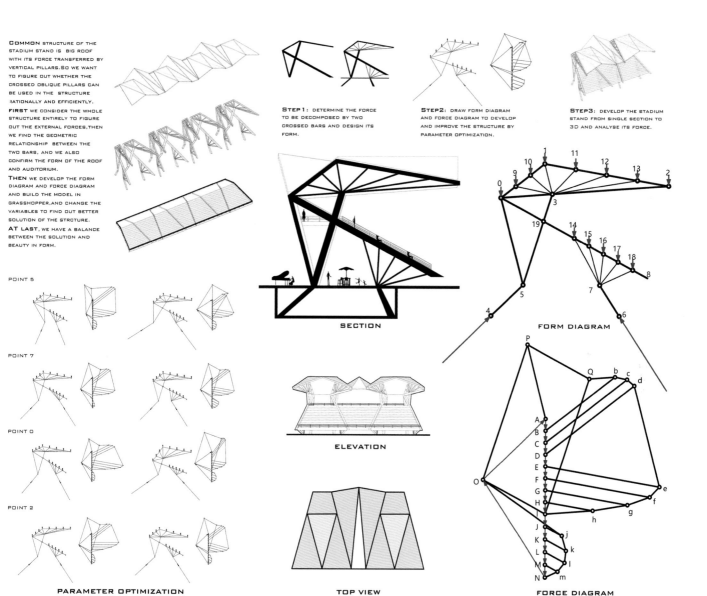

COMMON structure of the stadium stand is big roof with its force transferred by vertical pillars. So we want to figure out whether the crossed oblique pillars can be used in the structure rationally and efficiently.

FIRST we consider the whole structure entirely to figure out the external forces, then we find the geometric relationship between the two bars, and we also confirm the form of the roof and auditorium.

THEN we develop the form diagram and force diagram and build the model in grasshopper, and change the variables to find out better solution of the stroture.

AT LAST, we have a balance between the solution and beauty in form.

STEP1: determine the force to be decomposed by two crossed bars and design its form.

STEP2: draw form diagram and force diagram to develop and improve the structure by parameter optimization.

STEP3: develop the stadium stand from single section to 3D and analyse its force.

POINT 5

POINT 7

POINT 0

POINT 2

PARAMETER OPTIMIZATION

SECTION

ELEVATION

TOP VIEW

FORM DIAGRAM

FORCE DIAGRAM

ANALYSIS OF ROOF STRUCTURE

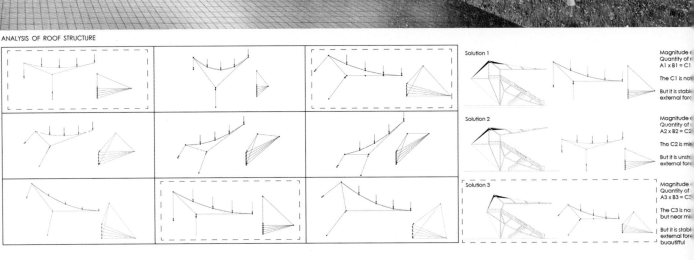

Solution 1

Magnitude o
Quantity of
A1 x B1 = C1

The C1 is not

But it is stable
external for

Solution 2

Magnitude o
Quantity of
A2 x B2 = C2

The C2 is mi

But it is unsto
external for

Solution 3

Magnitude
Quantity of
A3 x B3 = C3

The C3 is no
but near mi

But it is stabl
external for
beautiful

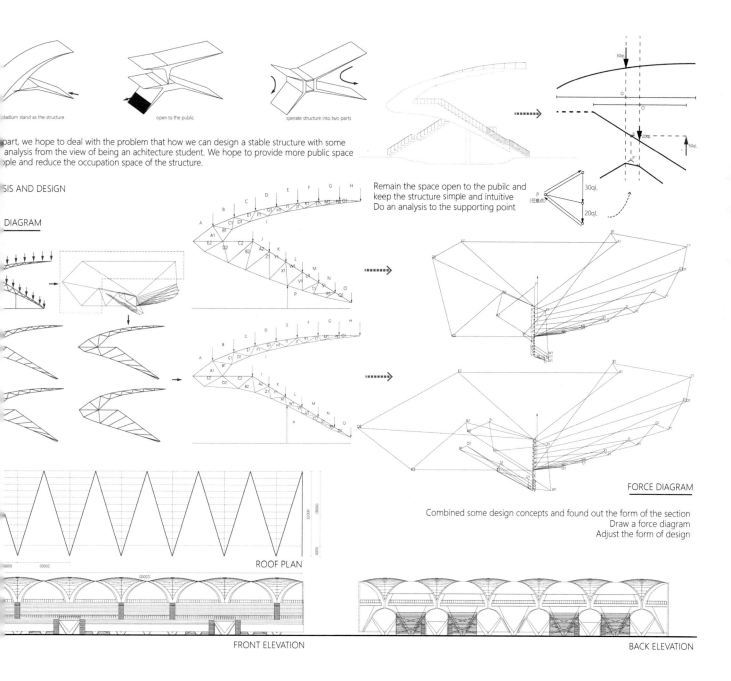

stadium stand as the structure

open to the public

seperate structure into two parts

part, we hope to deal with the problem that how we can design a stable structure with some
analysis from the view of being an achitecture student. We hope to provide more public space
ople and reduce the occupation space of the structure.

SIS AND DESIGN

DIAGRAM

Remain the space open to the pubilc and
keep the structure simple and intuitive
Do an analysis to the supporting point

FORCE DIAGRAM

Combined some design concepts and found out the form of the section
Draw a force diagram
Adjust the form of design

ROOF PLAN

FRONT ELEVATION

BACK ELEVATION

作为城市再生催化剂的预制功能单元

PREFABRICATED FUNCTIONAL UNIT AS URBAN REGENERATION CATALYST

罗伯特·博洛尼亚 华晓宁

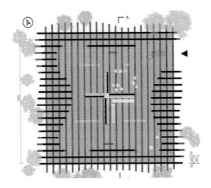

本课程通过对项目形式及其概念产生深远影响的技术知识，为学生介绍如何根据建筑原则，将建筑单元设计为居住空间和建筑空间。设计工作室关注项目的"建构"维度，基于所有学科的同等重要性和不同技巧之间的互补性，根据将建筑项目视为整个设计过程的概念，强调建筑方法、系统以及作为整个建筑表现的重要部分的材料特征。本课程为学生提供建筑设计的方法和操作工具，关注空间和技术系统以及与环境和背景的关系。考虑可持续性、预制、项目可行性、施工以及探索创新原则，建筑技术的方法和工具将指导设计过程。

本课程的主题是在城市环境中设计预制的最小功能单元。主题将根据作为项目背景的"UIA 霍普杯2019年国际大学生建筑设计竞赛"发展。出于教育目的，参加课程的学生将设计其提案，在小型降级城市区域中降低竞赛要求以实现微观建筑规模。学生将有机会在之后完成并提交项目以供竞赛，以便实践其教育经验并评估竞赛的方法。本课程鼓励学生探索"微观建筑和宏观设计"之间的维度，这一维度暗示着空间美学和功能、施工过程和产品设计之间的密切关系。

教学内容涉及：

1. 项目要求和功能空间方案
2. 预制方法和含义
3. 最小功能单元的建筑类型学
4. 技术资源

The course introduces students how to design a building unit according the principle of architecture as inhabited and built space through a technical knowledge that profoundly affect the form of the project and its conception.

The design studio focuses on the "tectonic" dimension of the project, emphasizes the building methods and system and the material character as significant part of the whole architectural expression, according to the concept of the architectu project as a whole design process based on the equivalence of all disciplines a complementarity between the different skills.

The course provides students with methodological and operational tools architectural design, focuses on both the spatial and technological systems and t relationship with environment and context. The methods and tools of architectu technology will guide the design process considering the principles of sustainabili prefabrication, project feasibility and construction as well as exploring innovation.

The theme of the course is the design of a prefabricated minimal functional unit an urban context. The theme will be developed according the "UIA HYP Cup 20 International Student Competition in Architectural Design" as a background of project. Students attending the course will design their proposal for educatio purposes and operate a downscaling of the competition requirement up to the mic architecture scale in a small degraded urban area. Students will have the opportur to complete and submit the project for competition at a later time in order to pu practice their educational experience and evaluate the contest's method. The cou encourages students to explore the dimension in between "micro-architecture a macro-design" which implies a close relationship among the aesthetic and functior the space, the construction process and the design of product.

Teaching contents deal with:

1. project requirements and functional-spatial programme
2. prefabrication methods and implications
3. building typology of a minimal functional unit
4. technological resource

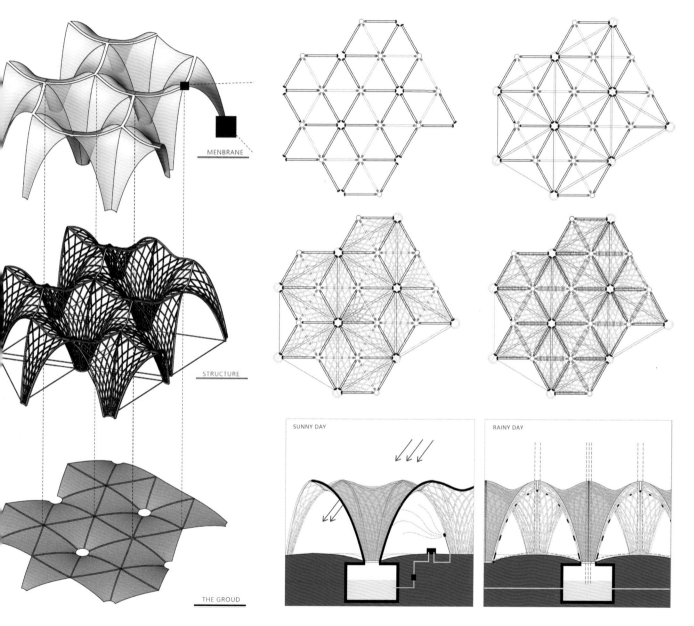

MENBRANE

STRUCTURE

THE GROUD

SUNNY DAY

RAINY DAY

CONCEPT AND FUNCTION

135

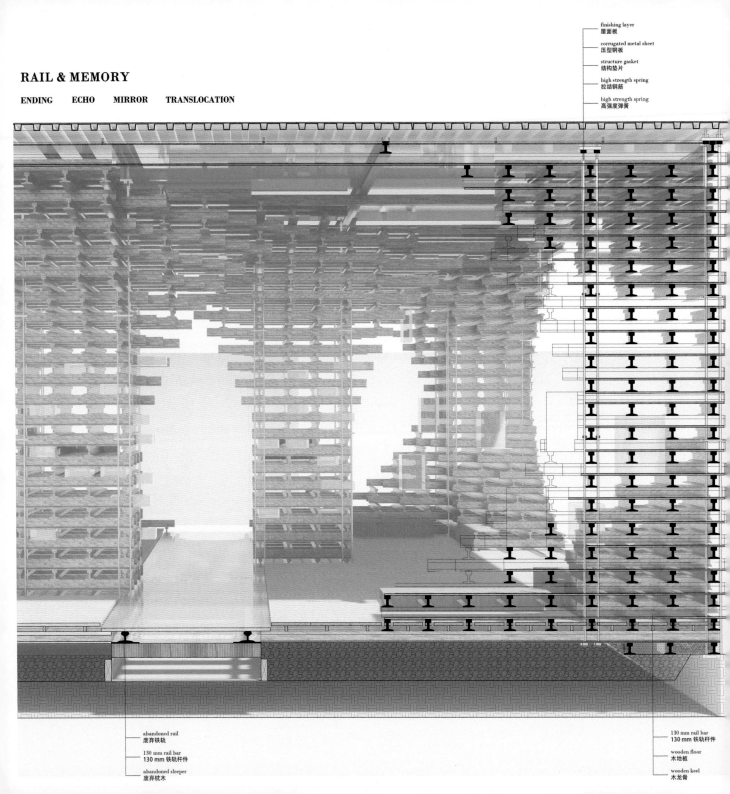

RAIL & MEMORY

ENDING   ECHO   MIRROR   TRANSLOCATION

finishing layer
屋面板

corrugated metal sheet
压型钢板

structure gasket
结构垫片

high strength spring
拉结钢筋

high strength spring
高强度弹簧

abandoned rail
废弃铁轨

130 mm rail bar
130 mm 铁轨杆件

abandoned sleeper
废弃枕木

130 mm rail bar
130 mm 铁轨杆件

wooden floor
木地板

wooden keel
木龙骨

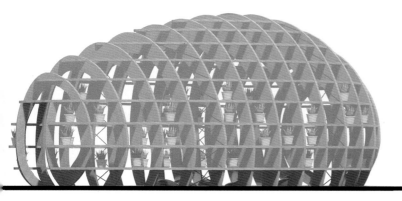

North Elevation 1:50

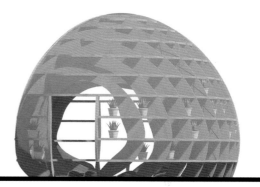

East Elevation 1:50

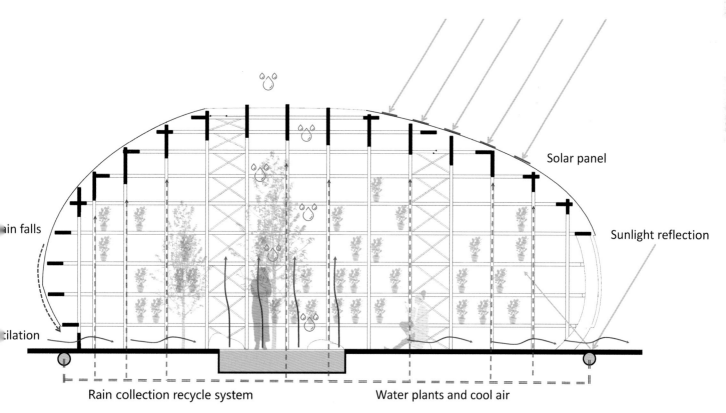

Solar panel

ain falls

Sunlight reflection

tilation

Rain collection recycle system

Water plants and cool air

137

使用3D打印技术的建筑生态节点设计

CONSTRUCTING HABITABLE STRUCTURES USING 3D PRINTED JOINERY

朴大权 钟华颖

　　"适应性节点"工作坊探索了3D打印节点的潜力，这些节点可采用新颖的方式连接传统材料（织物、木材/塑料/纸板、棒杆、网格和管道）。使用此节点设计，学生将设计并建造一个可居住构筑物（例如亭子、天篷、隔板或家具）。将介绍包括灵活性、不稳定性、适应性和可编程性在内的概念；并将研究柔性机构、双稳态机构、剪纸、折纸、软/柔性接头和混合接头等机构。

　　通过本工作坊，学生不仅能够发展数字设计和制造过程方面的技能和知识，而且能够学习如何在开发材料系统时应用高级几何学和筑造学。新兴的数字技术和过程，包括创成式设计、几何优化和设计脚本，将介绍为实践研讨会，学生将利用其开发自己的设计研究项目。

　　1.简介
　　2.案例研究展示
　　3.适应性节点的初步设计和原型
　　4.适应性节点的设计和原型制作
　　5.适应节点的变体和原型
　　6.聚合设计和原型展示
　　7.可居住构筑物的设计
　　8.可居住构筑物的建模
　　9.可居住构筑物的制造和文件1/3
　　10.可居住构筑物的制造和文件2/3
　　11.可居住构筑物的制造和文件3/3
　　12.最终审查

The "Adaptive Joints" workshop explores the potentials of 3D printed joints that ca connect conventional materials (fabric, wood/plastic/ paper boards, sticks, meshe and pipes) in novel ways. Using this joint design, the students will design an construct a habitable structure (e.g., pavilion, canopy, partition, or furniture). Concep including flexibility, instability, adaptability, and programmability will be introduce and mechanisms such as compliant mechanisms, bi-stable mechanisms, Kirigam Origami, soft/flexible joints, and hybrid joints will be investigated.

Through the workshop, the students will not only be able to develop skills ar knowledge of the digital design and fabrication process, but also learn how to app advanced geometry and tectonics in developing material systems. Emerging digit techniques and processes including generative design, geometry optimization, ar design scripting will be introduced as hands-on workshops, and the students w utilize them to develop their design research project.

　　1.Introduction
　　2.Case-study Presentation
　　3.Preliminary Adaptive Joint Design and Prototypes
　　4.Adaptive Joint Design and Prototyping
　　5.Adaptive Joint Variations and Prototype
　　6.Aggregation Design and Prototype Presentation
　　7.Habitable Structure Design
　　8.Habitable Structure Modeling
　　9.Habitable Structure Fabrication and Documentation 1/3
　　10.Habitable Structure Fabrication and Documentation 2/3
　　11.Habitable Structure Fabrication and Documentation 3/3
　　12.Final Review

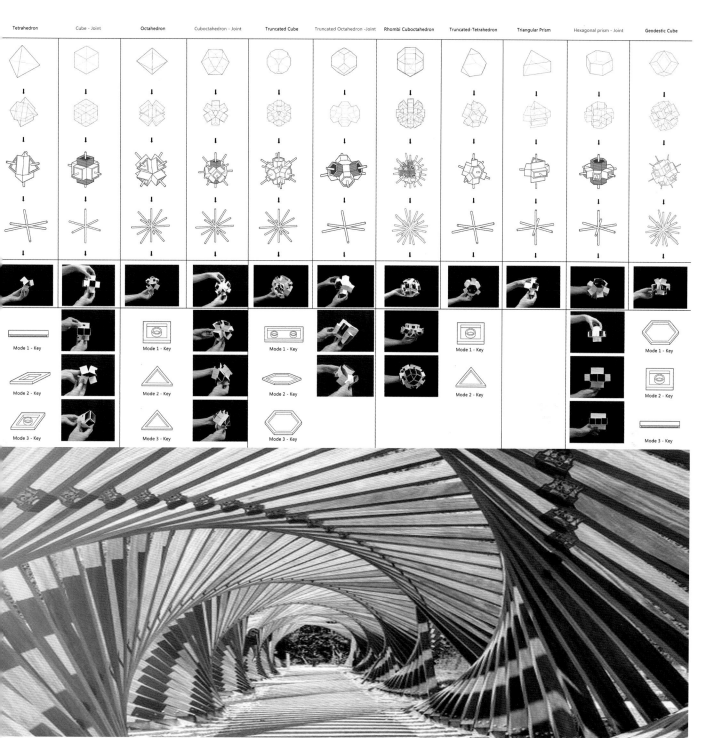

Tetrahedron	Cube - Joint	Octahedron	Cuboctahedron - Joint	Truncated Cube	Truncated Octahedron -Joint	Rhombi Cuboctahedron	Truncated-Tetrahedron	Triangular Prism	Hexagonal prism - Joint	Geodestic Cube

Mode 1 - Key   Mode 1 - Key   Mode 1 - Key   Mode 1 - Key   Mode 1 - Key   Mode 1 - Key

Mode 2 - Key   Mode 2 - Key   Mode 2 - Key   Mode 2 - Key   Mode 2 - Key

Mode 3 - Key   Mode 3 - Key   Mode 3 - Key   Mode 3 - Key

CASE STUDY

   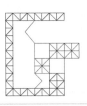 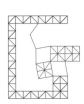

THE KEY POINT OF THE HANDLE IS THE COMPLIANT JOINT MAKE ONE MOVEMENT
TO ANOTHER MOVEMENT.

WE ANALYZE AND SIMPLIFY THE HANDLE TO A MOTION OF GEOMETRY.

PROTOTYPE RESEARCH

WE MAINLY STUDY THE MOTION OF TRIANGLES,
SQUARE AND PARALLELOGRAM, WE GET SOME PHYSICAL
PRINCIPLE IN IT.

PATTERN RESEARCH

   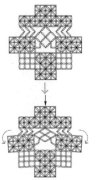

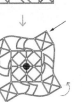 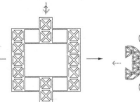 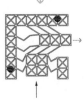

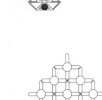

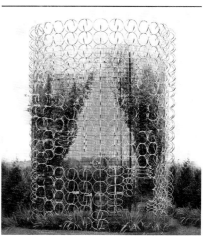

应对老龄化的混杂空间：建筑尺度
HYBRID SPACES OF NEGOTIATION FOR AN ACTIVE AGEING: MIDDLE-SMALL SCALE

米格尔·真蒂尔　傅筱

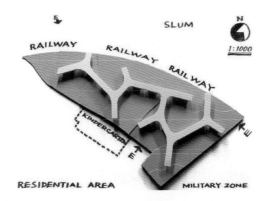

研讨会从大一中和中一小两个工作尺度探讨话题，日托中心是用于对混合空间开展实验的完美平台。私人和公共区域、内部和外部区域、自由行动和受控周边区域、私密和社交区域必须与邻里和建筑互动，以实现其在生活舒适度和治疗效果方面的潜力。从社会角度来看，老年人可与其他群体互动：学生、邻居、社会工作者、儿童，从而创造新的协同效应。城市和建筑布局能够在混合空间中呈现所有这些融合现实，并提出有助于形成用户活跃日常的空间关系。

学生将需要思考城市和建筑空间如何能够成为推动更具一体化的环境的催化剂，在此环境中，老年人将有机会享受变老的积极过程。换言之，在旨在增强横向社会互动的空间中开展身体和智力活动。学生的提案必须反映工地条件，并处理当地社会和传统的具体特征。该项目应为该话题考虑一个在所有意义上均可持续的方法：社会、生物气候、热力学、虚拟、生态等等。室内空间、花园区域、中间空间的设计应使其与彼此相互作用。

The workshop approaches the topic from two scales of work, big-middle and midd small, day-care center is a perfect platform for experimenting hybrid spaces. Priva and public, interior and exterior, freedom of movement and controlled perimete intimate and socialization areas must interact for the neighborhood and the buildi to achieve their potentiality in terms of living comfort and therapeutic effect. From social point of view, the elderly may interact with other groups: students, neighbo social workers, children, in order to create new synergies. Urban and architectu layouts are able to assume all these merging realities in hybrid spaces and propose spatial relationships that contribute to an active everyday of the users.

The students will be asked to reflect on the idea of how the urban and built spac are able to become catalysts towards a more integrative evironment, whe the elderly would have the chance of an active process of growing old. That developing activities, both physical and intellectual, in spaces designed to enhan transversal social interactions. Students' proposals must react to site conditio and deal with the specific features of local society and traditions. The project shou consider a sustainable approach to the topic, in every sense of it: social, bioclimat thermodynamic, virtual, ecological, etc.. Interior spaces, garden areas, intermedia spaces should be designed to interact with one another.

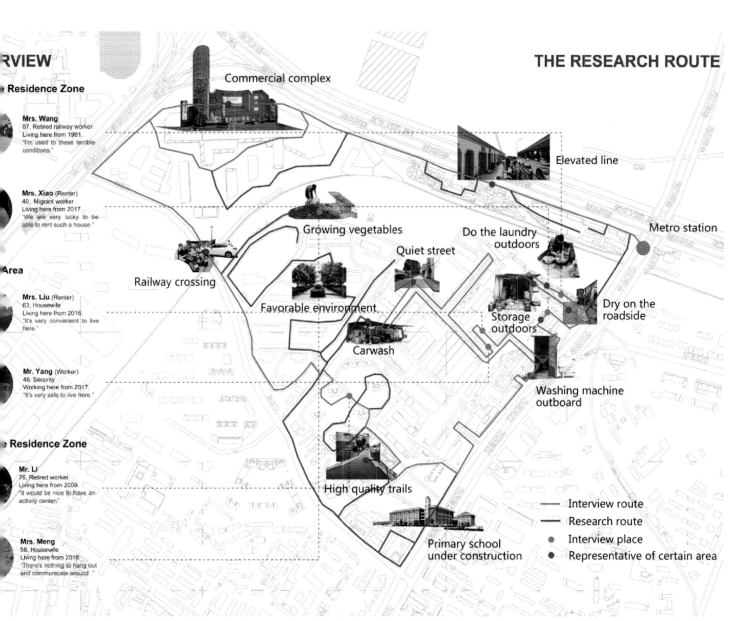

e Residence Zone

Mrs. Wang
87, Retired railway worker
Living here from 1961.
"I'm used to these terrible
conditions."

Mrs. Xiao (Renter)
40, Migrant worker
Living here from 2017.
"We are very lucky to be
able to rent such a house."

Area

Mrs. Liu (Renter)
63, Housewife
Living here from 2016.
"It's very convenient to live
here."

Mr. Yang (Worker)
48, Security
Working here from 2017.
"It's very safe to live here."

e Residence Zone

Mr. Li
76, Retired worker
Living here from 2009.
"It would be nice to have an
activity center."

Mrs. Meng
58, Housewife
Living here from 2016.
"There's nothing to hang out
and communicate around."

Commercial complex

Elevated line

Metro station

Growing vegetables

Do the laundry
outdoors

Quiet street

Dry on the
roadside

Railway crossing

Favorable environment

Storage
outdoors

Carwash

Washing machine
outboard

High quality trails

Primary school
under construction

—— Interview route

—— Research route

● Interview place

● Representative of certain area

143

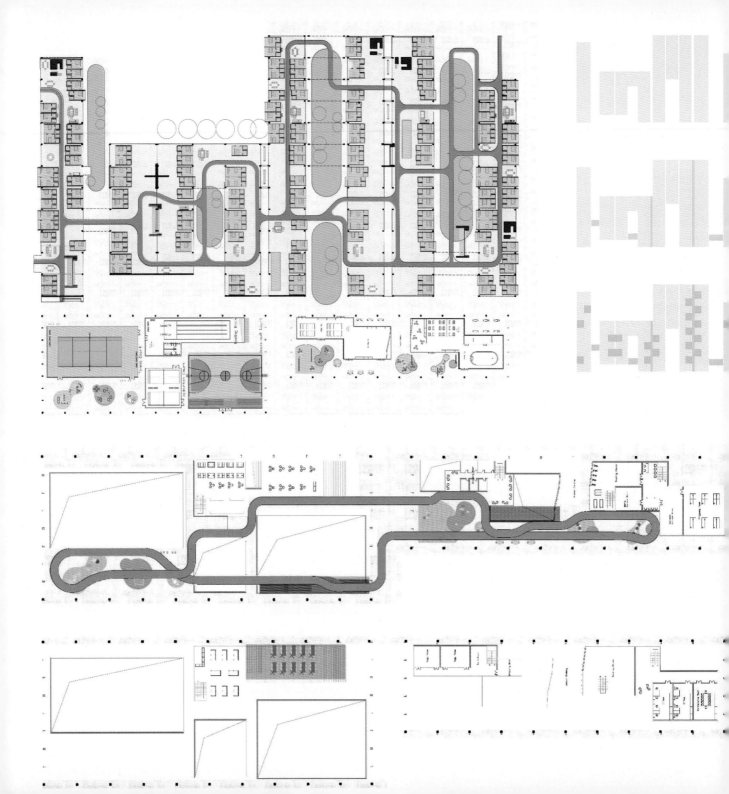

附录
APPENDIX

建筑设计课程
ARCHITECTURAL DESIGN COURSES

本科一年级
设计基础
·鲁安东　唐莲　尹航　黄华青
课程类型：必修
学时学分：64 学时／2 学分

Undergraduate Program 1st Year
BASIC DESIGN · LU Andong, TANG Lian,
YIN Hang, HUANG Huaqing
Type: Required Course
Study Period and Credits: 64 hours/2 credits

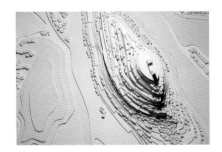

教学内容
　　课程采取了三个模块循序渐进、多个小组自主命题的新模式。三个模块分别为：A-感知（转化能力训练）、B-分析（制图能力训练）、C-创造（动手能力训练），旨在引导学生形成建成环境的初步认知，掌握建筑设计的初步技能。每个模块分为四个横向小组，由不同兴趣方向的教师在体系框架下充分发挥命题的自主性和创新性——包括人/环境、诗意/尺度等若干不同主题而又相互关联的子课程，由此编织成一张立体的"知识网格"。每个模块的教学时间为五周，各模块最后一周为交流评图。

Teaching Content
The course adopts a new model of three modules step by step and multiple group independent propositions. The three modules are A-perception (transformation ability training), B-analysis (cartography ability training) and C-creation (hands-on ability training). The purpose is to guide students to form a preliminary understanding of the built environment and master the preliminary skills of architectural design. Each module is divided into four horizontal groups. Teachers with different interest directions give full play to the autonomy and innovation of propositions under the framework of the system, including several sub-courses on different subjects such as human/environment, poetry/scale and so on, which weave into a three-dimensional "knowledge grid". The teaching time of each module is five weeks, and the last week of each module is to communicate and score charting.

本科二年级
建筑设计基础
·刘铨　冷天
课程类型：必修
学时学分：64 学时／4 学分

Undergraduate Program 2nd Year
BASIC OF ARCHITECTURAL DESIGN · LIU
Quan, LENG Tian
Type: Required Course
Study Period and Credits: 64 hours/4 credits

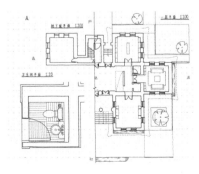

课题内容
　　认知与表达
教学目标
　　本课程是建筑学专业本科生的专业通识基础课程。本课程的任务主要是一方面让新生从专业的角度认知与实体建筑相关的基本知识，如主要建筑构件与材料、基本构造原理、空间尺度、建筑环境等知识；另一方面通过学习运用建筑学的专业表达方法来更好地掌握这些建筑基本知识，为今后深入的专业学习奠定基础。
教学内容
　　1.认知建筑
　　　　（1）立面局部测绘
　　　　（2）建筑平、剖面测绘
　　　　（3）建筑构件测绘
　　2.认知图示
　　　　（1）单体建筑图示认知
　　　　（2）建筑构件图示认知
　　3.认知环境
　　　　（1）街道空间认知
　　　　（2）建筑肌理类型认知
　　　　（3）地形与植被认知
　　4.专业建筑表达
　　　　（1）建筑图纸表达
　　　　（2）建筑模型表达
　　　　（3）环境分析图表达

Subject Content
Cognition and Presentation
Training Objective
The course is the basic course of general professional knowledge for undergraduates of architecture. The task of the course is, on the one hand, allow students to cognize basic knowledge about physical building from an professional perspective, such as main building components and materials, basic constructional principles, spatial dimensions, and building environment etc.; and on the other hand, to better master such basic architectural knowledge through studying application of professional presentation method of architecture, and to lay down solid foundation for future in-depth study of professional knowledge.
Teaching Content
1. Cognizing building
(1) Surveying and drawing of partial elevation
(2) Surveying and drawing plans, profiles of building
(3) Surveying and drawing building components
2. Cognizing drawings
(1) Cognition to drawings of individual building
(2) Cognition to drawings of building components
3.Cognizing environment
(1) Cognition to street space
(2) Cognition to types of building texture
(3) Cognition to terrain and vegetation
4. Professional architectural presentation
(1) Presentation with architectural drawings
(2) Presentation with architectural models
(3) Presentation with environmental analysis charts

本科二年级

建筑设计（一）：尺度与感知：独立居住空间设计 · 刘铨　冷天　王丹丹

课程类型：必修

学时学分：64学时 / 4学分

Undergraduate Program 2nd Year

ARCHITECTURAL DESIGN 1: SCALE AND PERCEPTION: DESIGN OF INDEPENDENT LIVING SPACE · LIU Quan, LENG Tian,WANG Dandan

Type: Required Course

Study Period and Credits: 64 hours/4 credits

课题内容

尺度与感知：独立居住空间设计

教学目标

本次练习的主要任务，是综合运用在建筑设计基础课程中的知识点，初步体验一个小型独立居住空间的设计过程。

训练的重点在于内部空间的整合性设计，同时希望学生在设计学习开始之初，能够主动去关注场地与界面、功能与空间、流线与出入口、尺度与感知等设计要素之间的紧密关系。

设计要点

1. 场地与界面：场地从外部限定了建筑空间的生成条件。作为第一个设计训练，教案对场地环境条件做了简化限定，主要是要求学生从场地原有界面出发来考虑新建建筑的形体、布局及其最终的空间视觉感受。

2. 功能与空间：使用者的不同功能需求是建筑空间生成的主要动因，也是建筑设计要解决的基本问题。本次设计的建筑功能为小型家庭独立式住宅并附设有书房功能，家庭主要成员包括一对年轻夫妇和一位未成年儿童（7岁左右），新建建筑面积 160~200 m²，建筑高度≤8 m。

3. 流线组织与出入口设置：建筑内部各功能空间需要合理的水平、垂直交通来相互沟通与联系；建筑的内部空间需要考虑与场地周边环境条件的合理衔接。

4. 尺度与感知：建筑内部的空间是供人来使用的，因此建筑中的各功能空间的尺度，都必须以人体作为基本的参照和考量，并结合人体的各种行为活动方式，来确定合理的建筑空间尺寸。

Subject Content

Scale and Perception: Design of Independent Living Space

Training Objective

The main task of this exercise is to comprehensively apply the knowledge points in the basic course of architectural design to preliminarily experience the design process of a small independent living space. The training focuses on the integrated design of internal space. At the beginning of design study, students are expected to take the initiative to pay attention to the close relationship between design elements as site and interface, function and space, streamline and access, scale and perception.

Key Points of Design

1. Site and interface: the site defines the generation condition of building space from the outside. As the first design training, the teaching plan simplifies and limits the environmental conditions of the site, mainly requiring students to consider the shape, layout and final spatial visual perception of new building from the original interface of the site. 2.Function and space: different functional requirements of users are the main motivation for the generation of building space and also the basic problem to be solved in architectural design. 3. Streamline organization and access setting: each functional space inside the building shall be mutually communicated and connected by rational horizontal and vertical traffic; the internal space of the building shall be properly connected with the surrounding environmental conditions of the site. 4. Scale and perception: the space inside the building is for use. Therefore, the scale of each functional space in the building must take human body as the basic reference and consideration, and combine with various behavioral and activity modes of human body to confirm rational architectural space size.

本科二年级

建筑设计（二）：空间与建构：小型多功能文化展廊设计 · 冷天　刘铨　王丹丹

课程类型：必修

学时学分：64学时 / 4学分

Undergraduate Program 2nd Year

ARCHITECTURAL DESIGN 2: SPACE AND CONSTRUCTION: DESIGN OF SMALL MULTI-FUNCTIONAL CULTURAL GALLERY · LENG Tian, LIU Quan, WANG Dandan

Type: Required Course

Study Period and Credits: 64 hours/4 credits

教学目标

本次练习为荷兰大使馆西侧辅助用房加建设计任务，要求综合运用在建筑设计基础课程中的知识点，初步操作一个小型公共建筑项目。

设计要点

1. 形体与场地：本案场地内部的荷兰大使馆旧址需要进行重点思考，包括新旧建筑的高度关系、体量关系和城市视线关系，对新建筑的形式都提出了具体的要求，必须从场地的体量与人的视点出发来考虑新建建筑的形体、布局、其形塑的城市建筑对话关系以及室外空间的活动氛围。

2. 空间与活动：本次设计的功能拟定为补充荷兰大使馆旧址在融入城市生活时所必不可少的公共空间，新建建筑面积不超过 300 m²，建筑高度≤5 m（檐口高度，不包括女儿墙）。

3. 服务与被服务：公共建筑与住宅的区别在于，活动规模和布置成为重中之重。建筑公共部分如何梳理组织成有秩序的部分？活动空间又如何与辅助空间形成良好的使用关系？流线、使用空间的方式、家具的布置，这些具体的建筑使用状态在开展活动时，建筑各部分之间的关系是什么样子的？需要我们细致思考并设定。

4. 结构与构造：从抽象的空间问题到实际的建造过程，一个好的建筑师，必须懂得结构和构造，才能变被动为主动，形成自己的实体建筑语言，成功操作建筑的实体部分实现自己的建筑目标。

Teaching Objective

This exercise is the design task of additional construction of auxiliary house in the west of Dutch embassy. Students are required to comprehensively apply knowledge points in the basic course of architectural design to preliminarily operate a small public building project.

Key Points of Design

1. Form and site: students shall focus on thinking of the former Dutch embassy site inside the site in this case, including the height relationship, volume relationship of new and old buildings and city sight relationship, and specific requirements are proposed on the form of new building.

2. Space and activity: the function of this design is proposed to supplement the public space necessary for the former Dutch embassy site to integrate into urban life. The new construction area is no more than 300 m², and the building height is ≤ 5 m (cornice height, excluding parapet wall).

3. Service and being served: How to organize the public part of the building into an orderly part? How to form a good use relationship between the activity space and the auxiliary space? Streamline, form of service space and the arrangement of furniture, what is the relationship between various parts of the building when the activity of these specific building service states is carried out? We need to think carefully and set.

4. Structure and construction: from the abstract space problem to the actual construction process, a good architect must understand structure and construction to change the passive into the active, form their own physical architectural language, and successfully operate the physical part of building to achieve their own architectural goals.

本科三年级
建筑设计（三）：幼儿园设计
· 童滋雨　华晓宁　钟华颖
课程类型：必修
学时学分：72学时／4学分

Undergraduate Program 3rd Year
ARCHITECTURAL DESIGN 3:
KINDERGARTEN DESIGN · TONG Ziyu,HUA
Xiaoning,ZHONG Huaying
Type: Required Course

Study Period and Credits: 72 hours/4 credits

课题内容
幼儿园设计
教学目标
此课程训练解决建筑设计中的一类典型问题：标准空间单元的重复和组合。建筑一般都是多个空间的组合，其中一类比较特殊的建筑，其主体是通过一些相同或相似的标准空间单元重复而成，这种连续且有规律的重复，很容易体现出一种韵律节奏感。对这类建筑的设计练习，可以帮助学生了解并熟悉空间组合中的重复、韵律、节奏、变化等操作手法。
教学内容
某幼儿园，用地面积约7200 m²。拟设小班、中班、大班各3个，共计9个班，每班人数为25人。使用面积约为2100 m²。高度不超过3层。

Subject Content
Kindergarten Design
Training Objective
This course is designed to solve a typical problem in architectural design: repetition and combination of standard space units. Architecture is generally a combination of multiple spaces, among which a special type of architecture is formed by repetition of some same or similar standard space units. Such continuous and regular repetition can easily reflect a sense of rhythm .The design exercise of such building can help students understand and be familiar with repetition, rhythm, change and other operational techniques in the spatial combination.
Teaching Content
A kindergarten, the land area is about 7200 m². It is planned to set up 3 junior classes, 3 middle classes and 3 senior classes, 9 classes in total, with 25 students in each class. The usable area is about 2100 m². The height is no more than 3 stories.

本科三年级
建筑设计（四）：文学博物馆
· 华晓宁　童滋雨　钟华颖
课程类型：必修
学时学分：72学时／4学分

Undergraduate Program 3rd Year
ARCHITECTURAL DESIGN 4: LITERATURE
MUSEUM · HUA Xiaoning, TONG Ziyu,ZHONG
Huaying
Type: Required Course

Study Period and Credits: 72 hours/4 credits

课题内容
文学博物馆
教学目标
本课程主题是"空间"，学习建筑空间组织的技巧和方法，训练空间的操作与表达。空间问题是建筑学的基本问题。课题基于复杂空间组织的训练和学习，从空间秩序入手，安排大空间与小空间、独立空间与重复空间，区分公共与私密空间、服务与被服务空间、开放与封闭空间。同时，充分考虑人在空间中的行为、空间感受，尝试以空间为手段表达特定的意义和氛围，最终形成一个完整的设计。
教学内容
本次设计拟在长江路历史文化街区建造一座以文学为主题的综合性博物馆，以促进社会文化事业的发展。基地位于南京长江路与中山东路之间、江苏美术馆新馆东侧原梅园中学地块，用地面积约7670 m²。新建筑应具备以下功能：
——展示南京古代与当代著名文学家、文学作品、文学活动的相关文物、资料；
——收藏与南京相关的文学资料、档案、文物；
——研究与南京相关的文学历史、理论、创作；
——文学普及和社会教育功能（如作家作品研讨会、签售会、文学沙龙、读者见面会等）。

Subject Content
Literature Museum
Training Objective
The subject of this course is "space", in which students learn the skills and methods of space organization in architecture, and operation and expression of training space. Space is the basic issue of architecture. The subject is based on the training and learning of complex space organization. Students shall start with the spatial order, arrange large space and small space, independent space and repeated space, distinguish public and private space, service and served space, open and closed space. At the same time, students shall full consider behavior of people and feelings in space, try to express specific meaning and atmosphere by means of space, and finally form a complete design.
Teaching Content
In this design, it is planned to build a comprehensive museum with the theme of literature in the historical and cultural block of Changjiang Road, so as to promote social and cultural development. The site is located between Changjiang Road and Zhongshan East Road in Nanjing and the former Meiyuan Middle School plot in the east of the new museum of Jiangsu Art Museum, with an area of about 7670 m². The new building should have the following functions:
—Display cultural relics, materials of Nanjing ancient and contemporary famous writers, literary works, literary activities;
—Collect literature materials, archives and cultural relics related to Nanjing;
—Study the history, theory and creation of literature related to Nanjing;
—Literature popularization and social education functions (such as writers' works seminar, book signing, literature salon, readers' meeting, etc.).

本科三年级
建筑设计（五）：大学生健身中心改扩建
设计 · 傅筱　钟华颖　王铠
课程类型：必修
学时学分：64 学时／4 学分

Undergraduate Program 3rd Year
ARCHITECTURAL DESIGN 5: RECONSTRUCTION
AND EXPANSION DESIGN OF COLLEGE STUDENT
FITNESS CENTER · FU Xiao, ZHONG Huaying,
WANG Kai
Type: Required Course
Study Period and Credits: 64 hours/4 credits

课题内容
大学生健身中心改扩建设计

教学目标
本课题以大学生健身中心为训练载体，学习并掌握中小
跨建筑的基本设计原理，掌握基本的结构类型与建筑形式空
间之间的逻辑关系，培养建筑结构、建筑空间与建筑设计的
协调能力。

教学内容
本项目拟在南京大学鼓楼校区体育馆基地处改扩建大学
生健身中心，以服务于南京大学师生，可适当考虑对周边
居民的服务。根据基地条件、功能使用进行建筑和场地设
计。基地用地面积9000 m²。现状基地由一座体育馆和一座
游泳馆（吕志和馆）组成，设计需保留体育馆，拆除现状
游泳馆并重新设计一座大学生健身中心，其建筑总面积约
4300 m²。建筑高度控制在 24 m 以下，注意场地东西向高
差，场地下挖不得超过一层，深度不超过 4.5 m。

Subject Content
Reconstruction and Expansion Design of College Student Fitness
Center
Training Objective
This subject takes college students' fitness center as the training
carrier, in which students learn and master the basic design
principles of small and medium-span building, master the logical
relationship between basic structural type and architectural form
space, and cultivate the coordination ability of architectural structure,
architectural space and architectural design.
Teaching Content
In this project, it is planned to renovate and expand the fitness center
for college students at the base of the gymnasium in Gulou Campus
of Nanjing University, so as to serve the teachers and students of
Nanjing University, and the service to surrounding residents can be
properly considered. Architecture and site are designed according to
the base condition and functional use. The base land area is 9000 m².
The current base consists of a gymnasium and a swimming pool (Lv
Zhihe Hall). The gymnasium shall be retained, the current swimming
pool shall be dismantled and a new fitness center for college
students shall be designed, with the total building area of about
4300 m². The height of the building shall be controlled below 24 m.
Students shall pay attention to the elevation difference between the
east and west of the site, dig no more than one floor below the site
and the depth shall not exceed 4.5 m.

本科三年级
建筑设计（六）：社区文化艺术中心
· 王铠　钟华颖　尹航
课程类型：必修
学时学分：64 学时／4 学分

Undergraduate Program 3rd Year
ARCHITECTURAL DESIGN 6: COMMUNITY
CULTURE AND ART CENTER · WANG Kai,
ZHONG Huaying, YIN Hang
Type: Required Course
Study Period and Credits: 64 hours/4 credits

课题内容
社区文化艺术中心

教学目标
本课题以社区文化艺术中心为训练载体，学习并掌握综
合功能建筑基本设计原理，理解老城区街区式建筑与城市环
境的逻辑关系，培养建筑结构、建筑空间与建筑功能的综合
组织能力。

教学内容
本项目拟在百子亭风貌区基地处新建社区文化中心，总
建筑面积约 8000 m²，项目不仅为周边居民文化基础设施服
务，同时也期望成为复兴老城的街区活力的文化地标。根据
基地条件、功能使用进行建筑和场地设计。总用地详见附
图，基地用地面积 4600 m²。

Subject Content
Community Culture and Art Center
Training Objective
This subject takes the community culture and art center
as the training carrier, in which students learn and master
the basic design principles of comprehensive functional
architecture, understand the logical relationship between
the block style architecture in the old urban area and the
urban environment, and cultivate the comprehensive
organizational ability of architectural structure, architectural
space and architectural function.
Teaching Content
In this project, it is planned to build a new community
cultural center at the base of Baiziting Historical and
Cultural Area, with total construction area of about 8000 m².
The project does not only serve the cultural infrastructure
of the surrounding residents, but is also expected to
become a cultural landmark to revive the vitality of the old
urban blocks. Architecture and site are designed according
to the base condition and functional use. Refer to the total
land use in the attached drawing. The land area of the
base is 4600 m².

本科四年级

建筑设计（七）：高层设计："未来摩天楼"

·周凌　胡友培　尹航

课程类型：必修

学时学分：64 学时／4 学分

Undergraduate Program 4th Year

**ARCHITECTURAL DESIGN 7: HIGH-RISE
DESIGN: "FUTURE SKYSCRAPER"** · ZHOU
Ling, HU Youpei, YIN Hang

Type: Required Course

Study Period and Credits: 64 hours/4 credits

课题内容

　　高层建筑设计："未来摩天楼"

教学目标

　　高层建筑设计是本科建筑课程中最复杂与综合的单体设计任务。本次课程设计希望学生掌握高层建筑的基本特点，研究当代高层建筑的设计策略，了解高层建筑涉及的相关规范与知识，提高综合分析并解决问题的能力。

教学内容

　　本次设计提供三个主要的高层建筑发展方向给学生进行研究：

　　1.结构驱动的性能化高层体型

　　2.景观驱动的绿色摩天楼

　　3.复合功能的立体城市

　　学生需选择一种发展方向进行深入研究，并将其作为设计策略运用在自己的设计中，充分发挥创造性，尝试设计一座"未来摩天楼"。

　　鼓励若干小组进行组际合作，组成大组共同进行场地研究，各自选取相邻地块进行设计，并在设计中考虑到地块之间的城市关系（包括场地交通、高层天际线、裙房与地下层连通等），可在最终图纸中考虑制作拼合总图和总体环境模型。

Subject Content

High-rise Design: "Future Skyscraper"

Training Objective

High-rise architectural design is the most complex and comprehensive single design task in undergraduate architecture courses.The design of this course expects students to master the basic characteristics of high-rise buildings, study the design strategies of contemporary high-rise buildings, understand relevant norms and knowledge of high-rise buildings, and improve the ability of comprehensive analysis and problem solving.

Teaching Content

This design provides three main development directions of high-rise buildings for students to study:

1. Structure-driven performance-driven high-rise body type

2. Landscape-driven green skyscraper

3. Three-dimensional city with complex functions

Students should choose a development direction for in-depth research and apply it as a design strategy in their own design, give full play to their creativity and try to design a "future skyscraper".

Several groups are encouraged to have cooperation and compose a large group for field research, respectively select adjacent plot for design, and consider urban relationship (including the transportation, high-rise skyline, podium and underground floor connection, etc), it can be considered in the final drawings to produce assembled general layout and overall environment model.

本科四年级

建筑设计（八）：城市设计

·丁沃沃　唐莲　尤伟

课程类型：必修

学时学分：64 学时／4 学分

Undergraduate Program 4th Year

**ARCHITECTURAL DESIGN 8: URBAN
DESIGN** · DING Wowo, TANG Lian, YOU Wei

Type: Required Course

Study Period and Credits: 64 hours/4 credits

课题内容

　　城市设计

教学内容

　　经历了三十多年快速城市化，我国沿海发达城市终于不得不结束轻松的扩张，开始面对土地终会枯竭的事实。基于我们身处的城市，探讨借由改变土地使用性质、增大建设密度和改善交通组织等方式来提高土地使用效率，既是我们城市设计相关学术探索的重要方向，同时也是我们基于教研平台持续进行的系列教学实验之一。在此背景下，本课程基于真实地块的设计训练，认知高密度城市空间形态的真正含义，了解城市建构角色和城市物质空间的本质和效能，初步掌握高质量城市空间和城市建筑组合之间的关系，同时，进一步深化空间设计的技能、方法与绘图能力。

教学目标

　　在设计开展之前，首先需要建立起对城市空间形态的准确理解，为此城市空间形态的认知训练与设计训练同等重要；另外，城市设计图示不同于常规的建筑学图示，也是训练的核心内容。

Subject Content

Urban Design

Teaching Content

After more than 30 years of rapid urbanization, the developed coastal cities in China have finally had to end their easy expansion and begun to face the fact that land will eventually be exhausted. Based on the city we live in, discussing to improve land use efficiency by changing land use nature, increasing building density and improving transportation organization is not only an important direction of academic exploration related to urban design, but also one of the series of teaching experiments we continue on the teaching and research platform. Under this background, the course bases on real plot design training, aims to make students understand the real meaning of high-density urban space form, know the essence and effect of the role of city building and urban physical space, preliminarily master the relationship between high quality urban space and urban architecture combination, at the same time, further deepen skill, method and drawing ability in space design .

Training Objective

To the senior students who contact urban design for the first time, they should first establish an accurate understanding of urban spatial form before starting design. Therefore, the cognitive training of urban spatial form is equally important as the design training. In addition, urban design diagrams are different from conventional architectural diagrams, which is also the core of the training

本科四年级

毕业设计

赵辰

课程类型：必修

学时学分：1 学期 /0.75 学分

Undergraduate Program 4th Year

GRADUATION PROJECT · ZHAO Chen

Type: Required Course

Study Period and Credits: 1 term /0.75 credit

课题内容

福建政和南京大学土木/营造研究基地设计及乐溪双湾复兴规划

教学内容

南京大学建筑与城市规划学院赵辰工作室，已经连续多年在此区域开展了规划设计等研究工作，随着工作的深入发展与地方政府酝酿在乐溪流域的双湾（长垄村）设立"研究基地"及"工作站"，进一步推动直接的乡村复兴工作。为此，需要进行建筑设计和乡村规划。

阶段一，现场调研与分析：南方传统村落的复兴计划；阶段二，专题研究，各单项建筑设计与专项研究。

教学目标

掌握建筑设计基本的技能与知识（测绘、建模、调研、分析），并能对特定的地域和历史建筑进行深入的设计研究（内容策划、建筑结构、构造），根据社会发展的需求，提出改造和创造的可能。在选定的村落之现状研究的基础上，进行村落景观空间的整体规划。并且，选择相关重点区域与建筑，进行专项的建筑设计。

Subject Content

Fujian Zhenghe Nanjing University Civil/Construction Research Base Design and Lexi Shuangwan Revitalization Planning

Teaching Content

Zhao Chen Studio of the School of Architecture and Urban Planning, Nanjing University has carried out research works as planning design, etc. for many years in this region. The in-depth progress and local government's planning to establish "research base" and "work station" in Shuangwan (Changlong Village) in the basin of Lexi further promote direct rural revival work. Therefore, architectural design and rural planning are required.

Stage 1, field survey and analysis: revival plan of southern traditional village;

Stage 2, special subject research, single architectural design and special research.

Training Objective

To master the basic skills and knowledge of architectural design (surveying and mapping, modeling, research and analysis), can conduct in-depth design research on specific regions and historical architecture (content planning, architectural structure, construction), and propose possibility of transformation and creation according to the requirements of social development. On the basis of studying the current situation of selected village, conduct overall planning of village landscape space. Moreover, select relevant key areas and buildings to carry out special architectural design.

本科四年级

毕业设计

丁沃沃

课程类型：必修

学时学分：1 学期 /0.75 学分

Undergraduate Program 4th Year

GRADUATION PROJECT · DING Wowo

Type: Required Course

Study Period and Credits: 1 term /0.75 credit

课题内容

江苏句容农科所更新：建筑利用与更新设计

教学内容

句容市是江苏的县级市，隶属于镇江，本项目位于句容市市中心。由于城市发展的需要，农科所将搬迁至郊外。农科所地块内不仅有原有与农业科学研究相关的大片绿地，还有作为本市的历史建筑被保留的原农科所办公建筑。地方政府决定利用现有地块现状和条件，增加城市公共活动空间，改善城市服务设施；同时保留历史记忆并适当提高城市用地效率，通过本地块的改造和利用在多个层面上为城市空间增值。

教学目标

我国已经进入城市化成熟期，城市扩张已逐渐停止，而城市更新将是建筑设计长期的任务。为此，本毕业设计将基于真实项目设置了城市更新的建筑设计题目，旨在通过真实项目的操作，训练学生如何面对复杂的真实问题，引入研究方法科学有效的调研与分析，找出关键问题，学会如何利用既有建筑、如何保存历史记忆、如何决策新建项目并完成设计，通过设计研究理解多重限定下的建筑设计、面向建造的真实问题和城市设计的现实意义。

Subject Content

Renewal of Jiangsu Jurong Institute of Agricultural Science: Architectural Utilization and Renewal Design

Teaching Content

Jurong City is a county-level city in Jiangsu, subordinate to Zhenjiang. This project is located in the center of Jurong City. Due to the requirement of urban development, the Institute of Agricultural Science will be moved to the outskirts. In the plot of the Institute of Agricultural Science, there is not only large-area green land related to research of agricultural science, and also the office building preserved as the historical building of this city. Local government decides to utilize the current situation and conditions of the existing land to increase urban public activity space and improve urban service facilities. At the same time, preserve the historical memory and the properly improve efficiency of urban land use, increase value of urban space on several levels through the transformation and utilization of this plot.

Training Objective

This graduation design will set the subject of architectural design based on urban renewal according to the real project, with the purpose of training students to face complex realistic problem through operating real project, introducing scientific and effective research and analysis, finding out key problem, leaning how to utilize existing buildings, how to preserve historical memory, how to make decision in new project and complete design, understanding architectural design under multiple limits, facing the realistic problem of construction and realistic meaning of urban design through design and research.

本科四年级
毕业设计
· 吉国华
课程类型：必修
学时学分：1 学期 /0.75 学分

Undergraduate Program 4th Year
GRADUATION PROJECT · JI Guohua
Type: Required Course
Study Period and Credits: 1 term /0.75 credit

课题内容

数字化设计与建造

教学内容

本课题以"创客空间"为主题，要求学生在学校自选环境中设计一处用地面积6 m×8 m，建筑面积为15～20m²的创客空间，以满足有创业目标的学生聚会、交流、创想、协作的功能需求，推进创新意识培养。课题着重要求从建筑设计的实际问题出发，用数字化的方法研究和解决问题，最终通过数控加工的方式来实现具有真实细节的构筑物。

教学目标

基于建筑数字技术，本毕业设计涵盖案例分析、设计研究以及建造实践等三个部分，旨在融合数字化设计与数控建造这两个过程，将传统建筑行业中分离的设计与营造用数字技术结合起来，完成从数字化设计到数字化建造的全过程。整个课程以实物模型为研究媒介，希望在形式生成与建造验证这一往复的过程中，引导学生逐步形成关联设计与建造过程的协同意识，以培养寻求物质逻辑合理性的主动思考，建立基于数字技术的建筑设计建造一体化思维。

Subject Content

Digital Design and Building

Teaching Content

This subject takes "maker space" as the theme, and require students to design a maker space with the land area of 6m×8 and building area of 15~20 m² in the self-selected environme in the school, so as to meet the functional needs of students w entrepreneurial goals to gather, communicate, create ideas a cooperate, and promote the cultivation of innovation consciousness

Training Objective

Based on the digital technology of construction, this graduati design covers case study analysis, design research and constructi practice, aims at integrating two processes of digital design a numerical control construction, combining separated design a construction in traditional construction industry by digital technolo and completing the whole process from digital design to dig construction. The whole course takes physical model as t research medium, hoping to guide students to gradually form sense of cooperation of related design and construction proce in the process of form generation and construction verification, as to cultivate active thinking for the rationality of material log and establish an integrated thinking of architectural design a construction based on digital technology.

本科四年级
毕业设计
· 周凌
课程类型：必修
学时学分：1 学期 /0.75 学分

Undergraduate Program 4th Year
GRADUATION PROJECT · ZHOU Ling
Type: Required Course
Study Period and Credits: 1 term /0.75 credit

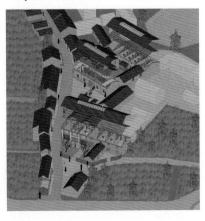

课题内容

未来城市与乡村：高密度人居环境设计研究

教学内容

第一，首先进行目标地区的城乡经济研究、城乡社会学研究、城乡地理研究；第二，进行现状调研，资料收集整理；第三，进行城乡聚落研究分析，建筑类型分析；第四，提出新的规划思想与策略；第五，完成概念规划；第六，完成类型填充与设计，完成重要地段建筑设计方案。通过案例研究、科学分析、实地调研，了解城乡高密度设计方法；通过几个阶段的学习研究，最终完成毕业设计。

教学目标

城乡融合发展与集约使用土地资源，对我国经济社会发展具有重要意义。芝加哥学派社会学家伯格斯提出城市同心圆理论，中国城乡也存在一种潜在的结构关系。课题通过调研，寻找城乡关系的内在结构，并且将其物化为规划建筑设计，完成从产业策划、战略规划到规划、建筑设计的完整知识产业链条。本次毕业设计将以科技部重点专项课题"东部发达地区传承中华建筑文脉的绿色建造体系"为依托，选择江苏、浙江、山东等东部地区的城市乡村，结合地方实际需要进行研究与规划设计，完成带有研究性和前瞻性的人居环境设计方案，提出一种中国东部地区城乡关系的优化模型。

Subject Content

Future City and Countryside : Research on the Design of Hig Density living Environment

Teaching Content

Firstly, research on urban and rural economy, urban and ru sociology and urban and rural geography in the target are secondly, research on current situation, data collection and collati thirdly, research and analysis of urban and rural settlement, analy of building type; fourthly, propose new planning ideas and strategi fifthly, complete the conceptual planning; sixthly, type filling a design, complete architectural design of important areas. Throu case study, scientific analysis and field research, understand t high-density design method of rural and urban areas; throu several stages of study and research, finally complete the graduat design.

Training Objective

Integrated development of urban and rural areas and intensive of land resources are of great significance to China's econom and social development. Chicago School sociologist Boggs forward the theory of concentric circles of cities, and there is a potential structural relationship between urban and rural areas China.Through research, this subject aims to seek for the inter structure of urban-rural relationship, materialize it into planning a architectural design, and complete the complete knowledge indust chain from industrial planning, strategic planning to planning a architectural design. This graduation design will depend on the l subject of Ministry of Science and Technology "Green Construct System of Eastern Developed Area to Inherit Chinese Architectu Culture", choose cities and country sides in Jiangsu, Zhejia Shandong, etc., combine with local actualities for research a planning design, and complete living environment design plan v the characteristics of research and foresight. An optimization mo of rural-urban relationship in eastern regions in China is proposed

本科四年级
毕业设计
华晓宁
课程类型：必修
学时学分：1 学期 /0.75 学分

Undergraduate Program 4th Year
GRADUATION PROJECT · HUA Xiaoning
Type: Required Course
Study Period and Credits: 1 term /0.75 credit

课题内容

　　"场所的时空与建筑的演变"：郎溪县定埠镇公共空间活化

教学内容

　　在中国广袤的土地上，正在发生着两种截然不同的变化：城市化的进程进一步加速，而乡村和小型城镇正在进一步的萎缩。城市化导致社会各要素的充分集中，集中过程中产生新的科技、人文、社交的需求和空间演变。乡村和城镇与之相反，各要素的流失导致老龄化、空心化，尤其是文化、教育等公共服务体系更是迅速瓦解。乡镇复兴并非仅仅意味着物质空间环境的重建或改造，更重要的是必须赋予乡镇生活以新的活力。

　　如何在新时代下构建缩小城乡差别的公共空间体系，结合场所的时空体现地域特色建筑的演变，创建新时代建筑本源人文特性和可持续演变模式，是本设计的主题。

教学目标

　　基于对郎溪县定埠镇历史和现状问题的调研分析，提出对镇区公共空间和相关节点建筑的活化更新设计方案。

Subject Content

"Time and Space of the Place and Architectural Evolution": Activation of Public Space in Dingbu Town, Langxi County

Teaching Content

Two distinct changes are taking place on the vast land of China: urbanization is further accelerated, while villages and small towns are further shrunk. Urbanization leads to the full concentration of social elements, and in the process of concentration, new demands on technology, culture and social contact and spatial evolution are generated. In rural areas, on the contrary, the loss of various elements leads to the aging and hollowing out, especially the rapid disintegration of public services such as culture and education. The revival of villages and towns does not only mean the reconstruction or reform of the material space environment, but more importantly, the life of villages and towns must be endowed with new vitality.

The theme of this design is how to build a public space system to narrow the difference between urban and rural areas in the new era, reflect the evolution of architecture with regional characteristics in combination with the space and time of the site, and create the original cultural characteristics and sustainable evolution mode of architecture in the new era.

Training Objective

Based on the research and analysis of the history and current situation of Dingbu Town in Langxi County, an activation and renewal design scheme for the public space and related node buildings in the town is proposed.

本科四年级
毕业设计
刘铨
课程类型：必修
学时学分：1 学期 /0.75 学分

Undergraduate Program 4th Year
GRADUATION PROJECT · LIU Quan
Type: Required Course
Study Period and Credits: 1 term /0.75 credit

课题内容

结合雨水综合利用的智慧教学综合体研究与设计

教学目标

　　1.研究探讨大学本科新教学模式对教学空间需求的变化。大学本科传统教学模式已不能适应当代高层次人才培养的需要。教室作为教学活动实施的主要场所，也必然面临改变。智慧教学空间需要通过"去教师中心化"及"教学与信息技术深度融合"，来支持和推动翻转课堂、研究性教学、个性化学习、学业评价及教学质量实时监督等教学改革的开展。在此过程中，教学空间的使用模式的变化，包括教室的布局、采光和声学要求、公共空间与教室配置关系等，都需要深入研究。

　　2.以雨水综合利用技术为核心的绿色建筑与景观设计方法推动"海绵城市"建设，在建筑与场地内提高雨水的生态化处理与利用是创造城市绿色生态城市的重要组成部分。在经济性、功能性与其他绿色节能要求限制条件下，如何更好地将绿色屋面、垂直绿化、景观绿色基础设施、人工雨水设施等结合到建筑空间与场地设计中，是需要大力研究的。

Subject Content

Research and Design of Intelligent Teaching Complex Combined with Comprehensive Utilization of Rainwater

Training Objective

1. Research and discuss the change of teaching space requirement, under the new teaching mode of undergraduate. Traditional teaching mode of undergraduate can no longer meet the needs of contemporary high-level talent training. As the main place for teaching activities, the classroom is bound to face changes. Intelligent teaching space shall support and promote the implementation of teaching reforms as flipped classroom, research-based teaching, personalized learning, academic evaluation and real-time supervision of teaching quality through "removing teacher centralization" and "deep integration of teaching and information technology". In this process, changes in the use mode of teaching space, including classroom layout, lighting and acoustic requirements, and the relationship between public space and classroom configuration, shall be further studied.

2. Green building and landscape design method with rainwater comprehensive utilization technology as the core promotes the construction of "sponge city". Improving ecological treatment and utilization of rainwater in buildings and sites is an important part of creating green and ecological city. Under the restriction of economy, function and other green energy saving requirements, how to better integrate green roof, vertical greening, green landscape infrastructure and artificial rainwater facilities into architectural space and site design shall be further studied.

本科四年级
毕业设计
· 童滋雨

课程类型：必修
学时学分：1 学期 /0.75 学分

Undergraduate Program 4th Year
GRADUATION PROJECT · TONG Ziyu
Type: Required Course
Study Period and Credits: 1 term /0.75 credit

课题内容
基于规则和算法的设计和搭建
教学内容
本题目聚焦于设计过程中的规则提取和算法应用，结合参数变量的设置，生成更有趣且合理的设计成果。在此过程中，设计的功能、形态乃至结构都可以是规则提取的目标。而相对于形式本身，我们更关注其潜在规则和算法的科学性。设计包括空间结构体的几何规则生成、参数化生成设计方法和设计算法等。另一方面，数字化加工技术的发展更增加了加工和建造的多样性，也为复杂建筑形体的实现提供了可能。目前可以使用的工具包括木工雕刻机、三维打印机和机器人系统。

教学目标
本题目初步设定是设计一个具有一定跨度和高度的构筑物，能够容纳5人左右的活动空间。要求空间适宜，结构合理，形体的生成应具有相应的几何规则和算法，并通过数字化加工完成最终的模型搭建。

Subject Content
Design and Construction Based on Rules and Algorithms
Teaching Content
This subject focuses on the rule extraction and algorithm applicatic in the design process, and generating more interesting and ration design results combined with the setting of parameter variables. this process, the function, shape and even structure of the desig can be the target of rule extraction. Compared with the form itse we pay more attention to the scientific nature of its underlyir rules and algorithms. Design includes geometric rule generatio parametric generation and design algorithm of space geometry. C the other hand, the development of digital processing technology ha increased the diversity of processing and construction, and provide the possibility for the realization of complex architectural forms. Toc currently available include woodworking engraving machines, 3 printers and robotic systems.

Training Objective
This subject is preliminarily set to design a structure with certa span and height, which can accommodate about 5 people. It required that the space is suitable, the structure is rational, t generation of form shall have corresponding geometric rules ar algorithms, and the final model building shall be completed throu digital processing.

本科四年级
毕业设计
· 王铠

课程类型：必修
学时学分：1 学期 /0.75 学分

Undergraduate Program 4th Year
GRADUATION PROJECT · WANG Kai
Type: Required Course
Study Period and Credits: 1 term /0.75 credit

课题内容
近郊乡村居住空间更新建筑工业化设计研究
教学内容
第一，项目背景研究，提出问题；第二，项目策划及案例研究（产品案例、工业化技术案例），解决问题的可行路径选择；第三，土地利用、在地建造工业化技术方案、产品选型等的综合研究，生成概念规划；第四，深化完成规划设计方案、典型建筑设计方案及其技术节点的深化。通过几个阶段的学习研究，最终完成毕业设计。

教学目标
空心村是我国当下乡村衰落的普遍现实，而乡土聚落的人居环境价值，是乡村被严重低估的方面。相对于城市，乡村的宜居特性，被基础设施的滞后、消费文化的冲击所掩盖。设计选取南京近郊某乡村，80～100户规模的乡土聚落，围绕居住空间更新的目标，以建筑工业化为主题，进行规划+建筑+产品设计的系统化设计研究。

Subject Content
Study on Industrial Design of Residential Space Renewal in Ru Areas
Teaching Content
Firstly, project background research, propose question; second project planning and case study (product case, industri technology case), select feasible path to solve the problem; third comprehensive research of land use, on-site construction industr technical scheme, product type selection, generate conceptu planning; fourthly, deepen and complete planning and desi scheme, typical architectural design scheme and the deepenir of technical nodes. Through several stages of study and researc finally complete the graduation design.

Training Objective
Hollow village is the universal reality of rural decline in China, a the value of living environment of rural settlements is serious underestimated. Compared with cities, the livable characteristics rural areas are covered by the lag of infrastructure and the impa of consumer culture. The design selects a rural village in the subu of Nanjing, a rural settlement of 80 ~ 100 households, and carri out systematic design research on planning + architecture + prodi design according to the objective of residential space renewal a the theme of architecture industrialization.

本科四年级
毕业设计
尹航
课程类型：必修
学时学分：1 学期 /0.75 学分

Undergraduate Program 4th Year
GRADUATION PROJECT · YIN Hang
Type: Required Course
Study Period and Credits: 1 term /0.75 credit

课题内容

林地服务中心建筑设计及周边景观规划

教学内容

本次设计依托真实地形，在实际建筑设计任务之外增加周边山林地景观规划设计。规划基地位于南京市江北新区老山余脉，为低山丘陵地区。基地周边有大片山林地和寺庙、石刻遗迹等，自然与人文资源丰富。但附近也有已关闭的矿坑和厂区等工业废弃地。本项目旨在在进行林区保护和矿区修复的同时，以林区自然保护和当地人文资源为主题，设计一座为林区保护提供一定办公空间，同时提供一定的教育培训、青少年活动基地、住宿、餐饮、会议等接待功能的服务中心建筑。

教学目标

本次设计围绕林区服务中心进行景观规划和建筑设计。服务中心可以集中布置，亦可以建筑组团的形式出现。功能可包含林区保护管理、乡风文化展示、教育、住宿、餐饮、会议中心等。在保证基本功能外，可根据情况设置其他相关功能，具体功能面积比例自行研究决定。周边景观规划需对工业废弃地进行交通和场地改造，对林地可进行一定范围的设计调整。

Subject Content

Architectural Design and Surrounding Landscape Planning of Woodland Service Center

Teaching Content

This design depends on the real terrain, and adds surrounding mountain and forest landscape planning design besides the actual architectural design task. The planning base is located in ranges of Old Mountain in Jiangbei New Area, Nanjing, as the low mountain hilly area. The base is surrounded by a large area of forest, temples and stone carving relics, rich in natural and cultural resources. However, there are also closed mines and industrial wastelands nearby. This project aims to design a service center building themed in forest area protection and local cultural resources, providing certain office space for forest protection, with the functions of education and training, youth activity base, accommodation, catering, conference, etc., while carrying out forest protection and mining restoration.

Training Objective

The design aims to have land planning and architectural design for the forest service center. The service center can be centrally arranged or in the form of architectural groups. The functions can include forest protection and management, rural culture display, education, accommodation, catering, conference center and so on. In addition to the basic functions, other relevant functions can be set according to the situation, and specific function area proportion can be independently researched and decided. The surrounding landscape planning shall have transportation and site transformation on the abandoned industrial land, and the forest land can have design adjustment in a certain scope.

研究生一年级
建筑设计研究（一）：基本设计
张雷
课程类型：必修
学时学分：40 学时 / 2 学分

Graduate Program 1st Year
DESIGN STUDIO 1: BASIC DESIGN · ZHANG Lei
Type: Required Course
Study Period and Credits: 40 hours/2 credits

课题内容

传统乡村聚落复兴研究

教学目标

课程从"环境""空间""场所"与"建造"等基本的建筑问题出发，对乡村聚落肌理、建筑类型及其生活方式进行分析研究，通过功能置换后的空间再利用，从建筑与基地、空间与活动、材料与实施等关系入手，强化设计问题的分析，强调准确的专业性表达。通过设计训练，达到对地域文化以及建筑设计过程与方法的基本认识与理解。

研究主题

乡土聚落 / 民居类型 / 空间再利用 / 建筑更新 / 建造逻辑

教学内容

对选定的乡村聚落进行调研，研究功能置换和整修改造的方法和策略，促进乡村传统村落的复兴。

Subject Content

Research on Revitalization of Traditional Rural Settlements

Training Objective

This course starts with basic architectural problems such as "environment" "space" "place" and "construction", analyzes and studies the texture, architectural type and life style of rural settlement, strengthens the analysis of design problems and emphasizes accurate professional expression from the relationship between building and base, space and activities, and material and implementation through spatial reuse after function replacement, to obtain basic knowledge and understanding of the regional culture as well as the process and methods of architectural design.

Research Subject

Rural settlement / types of folk house / reutilization of space /building renovation / constructional logic

Teaching Content

Conduct investigation and research on selected rural settlement, study methodology and strategy of function replacement and renovation and improvement, and promote revitalization of traditional rural villages.

研究生一年级
建筑设计研究（一）：基本设计
· 傅筱
课程类型：必修
学时学分：40 学时／2 学分

Graduate Program 1st Year
DESIGN STUDIO 1: BASIC DESIGN · FU Xiao
Type: Required Course
Study Period and Credits: 40 hours/2 credits

课题内容
宅基地住宅设计
教学目标
课程从"场地、空间、功能、经济性"等建筑的基本问题出发，通过宅基地住宅设计，训练学生对建筑逻辑性的认知，并让学生理解有品质的设计是以基本问题为基础的。
研究主题
设计的逻辑思维
教学内容
在 A、B 两块宅基地内任选一块进行住宅设计。

Subject Content
Homestead Housing Design
Training Objective
The course starts from fundamental issues of architecture such as "site, space, function, and economy", aims to train students to cognize architectural logics, and allow them to understand the quality design is based on such fundamental issues.
Research Subject
Logical Thinking of Design
Teaching Content
Select one from two homesteads A and B, conduct housing design

---

研究生一年级
建筑设计研究（一）：概念设计
· 梅尔加 · G. 麦尔格
课程类型：必修
学时学分：40 学时／2 学分

Graduate Program 1st Year
DESIGN STUDIO 1: CONCEPTUAL DESIGN · Sergio G. MELGAR
Type: Required Course
Study Period and Credits: 40 hours/2 credits

课题内容
建筑技术设计：中国绿色建筑新策略
教学目标
从遗迹到当代建筑，寻求培养新建筑师以成功应对当代社会对建筑专业人员提出的新挑战。富有创造力的建筑师，能够更好地协商建筑方案，能够在设计、最佳室内舒适度和建筑环境性能之间找到一致性。
教学内容
本课程的目的集中在以下六个方面：
1.遗迹干预：现有建筑、当地战略、古代工艺等
2.被动设计：形式、空间、功能、程序、日照设计、自然照明、热外壳、气密外壳等
3.主动设施：空气更新、冷却/供暖、热水供应、照明等
4.能源改造：现有建筑特征、被动策略、主动策略等
5.可再生综合能源：地热、光伏、风力、燃料电池等
6.仪表和控件：温度、湿度、二氧化碳浓度、能量、风速、风向、辐照度等
7.能效研究：科学数据库、研究论文、研究项目等

Subject Content
Architectural Technology Design: New Green Building Strategies China
Training Objective
From heritage to contemporary architecture, pursues the training of new architects to face successfully the new challenges posed by contemporary society to architecture professionals. Creative architects have an improved ability to negotiate an architectural proposal capable of finding coherence between the design, the best interior comfort degree and the environmental performance of the building.
Teaching Content
The objectives of the course are focused in six topics:
1.Heritage intervention: existing buildings, vernacular strategies, ancient technics, etc.
2. Passive design: form, space, function, program, solar design, natural lighting, thermal envelope, airtight envelope, etc.
3.Active facilities: air renovation, cooling/heating, hot water supply, lighting, etc.
4.Energy retrofitting: existing building characterization, passive strategies, active strategies, etc.
5.Renewable integrated energy: geothermal, photovoltaic, wind power, fuel cells, etc.
6.Instrumentation and control: temperature, humidity, C02 ppm, energy, wind speed, wind direction, irradiance, etc.
7.Energy efficiency research: scientific databases, research papers, research projects, etc.

研究生一年级
建筑设计研究（一）：概念设计
· 周凌
课程类型：必修
学时学分：40 学时／2 学分

Graduate Program 1st Year
DESIGN STUDIO 1: CONCEPTUAL
DESIGN · ZHOU Ling
Type: Required Course
Study Period and Credits: 40 hours/2 credits

课题内容
垂直城市：一种高密度建筑集群设计研究
教学目标
本课程要求在南京鼓楼滨江下关 CBD 地块，设计一座垂直城市，主要功能为居住、办公、酒店等，辅助功能自行定义，高度150～200 m。 设计既要有很强的超前概念，又要有具体的技术措施和实施策略，要具有现实可操作性，包含详细的功能和平面细化。最终提供一个面对现实问题、可以实际实施的建筑方案，完成一次具有高度职业化特征的设计研究训练。
教学内容
"垂直城市"指一种能将城市要素包括居住、工作、生活、休闲、医疗、教育等一起装进一个或几个体量里的巨型建筑。在"垂直城市"里，可以提供完备的城市居住、工作和生活功能，如银行、邮局、餐饮、购物、办公等。"垂直城市"拥有庞大的体量、超高的容积率、惊人的高度、少量的占地、高密度的居住人群等特征。"垂直城市"对土地的极度高效利用，可以释放更多空间给城市绿地和居住空间，保证生态绿地的规模化。

Subject Content
Vertical City : Research on Design of a High-density Building Cluster
Training Objective
This course requires the design of a vertical city in CBD plot of Binjiang Xiaguan in Gulou District, Nanjing. The main functions are residential, office, hotel, etc.. Auxiliary functions are self-defined and the height is 150~200 m. The design shall not only have a strong advanced concept, but also have specific technical measures and implementation strategies, to be practical and operable, including detailed functions and plane refinement. Finally, an architectural scheme that can be implemented in face of practical problems shall be provided, and a design research training with highly professional characteristics shall be completed.
Teaching Content
"Vertical city" refers to a huge building which can combine the urban elements including dwelling, working, living, leisure, medical care, education, etc. into one or several volumes. In "vertical city", complete urban dwelling, working and living functions, such as bank, post office, catering, shopping and office can be provided. "Vertical city" has huge volume, super high plot ratio, amazing height, less land cover, high density of residents, etc.. The extremely efficient use of land in "vertical city" can release more space for urban green space and living space, and ensure the scale of ecological green space.

研究生一年级
建筑设计研究（一）：概念设计
· 鲁安东
课程类型：必修
学时学分：40 学时／2 学分

Graduate Program 1st Year
DESIGN STUDIO 1: CONCEPTUAL
DESIGN · LU Andong
Type: Required Course
Study Period and Credits: 40 hours/2 credits

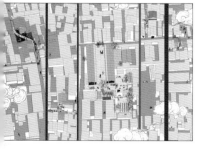

课题内容
面向未来的场所设计
研究主题
当代新兴的技术与社会条件改变了传统的空间规则。以时空一体化为基础的身体体验只以碎片的形式存在。新的空间规则需要依赖新的经验模型。新的经验模型包括但不限于"沉浸"与"交互"。 新的经验模型使得空间不再是感知连续的。在新的经验模型中，"内部性"空间崛起，"外部性"空间瓦解或者转化为另一层级的"内部性"。"场所"成为新的空间体系中的内核。"场所"的呈现需要通过多种媒介的合成。
教学内容
对当代场所进行类型特征解析（例如分散式、映射、分体、层叠、移动……）。对当代场所类型的历史演变进行研究。选取一目标场所的合成中介进行分析（带Scenario的任务书设定）。技术研发（与智能科技公司合作完成）。整合设计与表现。

Subject Content
Future-oriented Site Design
Research Subject
Contemporary emerging technology and social conditions have changed the traditional space rules. Physical experiences based on the integration of time and space exist only in fragments. New spatial rules shall depend on new empirical models. New empirical model includes, but is not limited to, "immersion" and "interaction." New empirical model makes space no longer perceptual continuously. In the new empirical model, the "internal" space rises, and the "external" space collapses or transforms into another level of "internal". "Place" becomes the core of new space system. The presentation of "place" requires the synthesis of multiple media.
Teaching Content
Typological analysis of contemporary sites (e.g., distributed, mapped, split, layered, mobile...). Research on the historical evolution of contemporary site types. Select a targeted site for synthesis mediation analysis (assignment setting with Scenario). Technology research and development (completed in cooperation with intelligent technology company). Integrate design and presentation.

研究生一年级
建筑设计研究（二）：建构研究
·尤多·特恩尼森
课程类型：必修
学时学分：40 学时／2 学分

Graduate Program 1st Year
DESIGN STUDIO 2: TECTONIC STUDY · Udo THEONNISSEN
Type: Required Course
Study Period and Credits: 40 hours/2 credits

课题内容

木构研究

教学目标

以中国木结构的悠久历史为背景，探讨传统和创新技术如何在城市环境中创造新的木结构形式。多功能建筑的建筑类型用于体验木结构的不同原理、构筑物和体积的关系以及一般和特定空间的作用。基于中国城市的快速发展和重建，我们还希望考虑物流和施工过程。人们经常忘记城市中的建筑是一种临时的材料商店。如果施工明智地使用资源，并考虑拆卸情况，则其具有可持续性。

教学内容

第一步将研究不同的传统、最佳实践和推测性建筑原则，从单个接头开始，演变成多层建筑系统。如何通过单一元素来表达建筑，将产生哪些空间后果？第二步将介绍工地，然后介绍建筑程序。每周一次的简短讲座将提供基本知识，包括材料特性、结构性能、制造、防火和隔音。在设计过程中，实体模型在理解试验过程、材料和形式间的相互关系方面发挥着重要作用。

Subject Content
Workshop Wood Construction
Research Objective
Against the background of the rich history of timber construction in China, explore how traditional and innovative techniques could create new forms of wood constructions in urban context. The building type of the multi-functional building serves to experience different principles of timber construction, the relation of structure and volume, and the role of generic and specific spaces. Based on the rapid development and reconstruction of the Chinese city, we also want to think about the material flows and the construction process. It is often forgotten that the buildings in the city are a kind of temporary material store. A construction is sustainable if it uses resources intelligently and if also disassembly is taken into account.
Teaching Content
In the first step, different traditional, best-practice and speculative construction principles will be studied, starting from a single joint and evolving to a multi-storey construction system. How is the architectural expression informed by the single element and which spatial consequences are arising? In the second step, the site will be introduced, after that the architectural program follows. In short weekly lectures the basic knowledge will be provided including material properties, structural performance, manufacturing, fire and acoustic protection. During the design course physical models play a substantial role in order to understand the interrelation of the triad process, material and form.

研究生一年级
建筑设计研究（二）：城市设计
·鲁安东
课程类型：必修
学时学分：40 学时／2 学分

Graduate Program 1st Year
DESIGN STUDIO 2 : URBAN DESIGN · LU Andong
Type: Required Course
Study Period and Credits: 40 hours/2 credits

课题内容

基于交互场所的城市设计

研究主题

当代新兴的技术与社会条件改变了传统的空间规则。不仅城市场所本身的交互形式在发生变化，同时在场所群组之间也存在着全新的空间关系和经验类型。城市设计需要为近期未来的城市生活提供结构性的支持，并激励新的空间创新、环境创新和技术创新。如何基于这一新的前提条件开展城市设计是本课程的目标。

研究内容

本课程以"颐和路历史文化街区保护利用"项目为载体，开展针对性的城市设计研究。颐和路历史文化街区源自民国"首都计划"中的新住宅区第一区，包括280余幢高级住宅及使领馆，集中呈现了民国的城市文化、历史政治、物质环境等。如何将丰富的叙事资源与先进的智慧技术相结合，使颐和路历史文化街区转变为南京的"城市空间实验室"是本项目的主要研究内容。

Subject Content
Urban Design Based on Interactive Place
Research Subject
Contemporary emerging technology and social conditions have changed the traditional space rules. Not only the interactive forms of urban places themselves are changing, but also new spatial relations and experience types exist among groups of places. Urban design shall provide structural support for urban life in the near future and stimulate new spatial, environmental and technological innovation. How to develop urban design based on this new premise is the objective of this course.
Teaching Content
This course will take "Protection and Utilization of Yihe Road Historic and Cultural Block" as the carrier and carry out targeted urban design research. Yihe Road Historic and Cultural Block originated from the first area of the new residential area in the Capital Plan of the Republic of China, including more than 280 high-grade residential buildings and embassies and consulates, which reflect the urban culture, historical politics and material environment of the Republic of China. The main research content of this project is how to combine rich narrative resources with advanced intelligent technology to transform Yihe Road Historic and Cultural Block into an "urban space laboratory" in Nanjing.

建筑设计研究（二）：城市设计

胡友培

课程类型：必修

学时学分：40 学时／ 2 学分

DESIGN STUDIO 2: URBAN DESIGN

HU Youpei

Type: Required Course

Study Period and Credits: 40 hours/2 credits

课题内容

　　城市再生：内边缘转型计划

教学目标

　　场地研究——在南京都市区内，选择一处场地，规模不限，作为设计场地。研究场地在城市化进程中的变迁，阐明、呈现场地作为内边缘的独特性。

　　场地愿景——从城市发展的全局视角，研判场地的价值与潜力，为场地建构新的身份提出新的愿景，制订场地的功能计划与策略性的项目介入，以激发场地的转型。

　　原形设计——通过对建筑、基础设施、景观以及它们的混杂，开展原形设计，赋予场地某种适宜的、具有想象力的空间形态。

　　在城市内部的自留地上，进行一场异托邦的建筑学实验与冒险。

Subject Content

Urban Regeneration: Transforming the Inner Edges

Teaching Objective

Site study — in the urban area of Nanjing, select a site of unlimited size as the design site. Study the changes of the site in the process of urbanization, clarify and present the uniqueness of the site as the inner edge.

Site vision — from the overall perspective of urban development, evaluate the value and potential of the site, construct a new identity for the site, put forward a new vision, formulate the functional plan of the site, and strategic project intervention to stimulate the transformation of the site.

Prototype design — develop prototype design through architecture, infrastructure, landscape and their mixture, endow the site with some appropriate and imaginative spatial form.

An architectural experiment and adventure of heterotopia is carried out on the private plot within the city.

建筑设计研究（二）：城市设计

凯里·思莱斯

课程类型：必修

学时学分：40 学时／ 2 学分

DESIGN STUDIO 2: URBAN DESIGN

Cary SIRESS

Type: Required Course

Study Period and Credits: 40 hours/2 credits

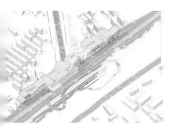

课题内容

　　异质类型：建筑、基础设施和地景

教学目标

　　本课程旨在理解城市建筑与城市物理空间的作用本质，初步把握建筑、城市空间与文化环境之间的关系，通过城市环境设计实践进一步理解空间设计的技巧和方法。

教学内容

　　设计工作室引入混合城市类型学主题，该主题借鉴建筑的空间制造能力、基础设施的服务分级性能和景观设计的地域组成，从而增强21世纪城市设计政策和实践的能力。设计课程为全球城市化的当代动态提供一系列重要设计视角，并尤其关注全球城市化过程对当地建筑和居住空间的当地影响。通过借鉴建筑、土木工程和景观设计的不同学科专有技术，城市设计重新构造为不单单为广大消费者提供标准空间产品的事务。

Subject Content

Heterogeneous Types: Architecture, Infrastructure & Landscape

Teaching Objective

This course aims to understand the nature of roles of urban buildings and urban physical space, and preliminarily grasp the relationships between architecture, urban space and cultural environment, to further understand skills and methodology of space design through practices of urban environmental design.

Teaching Content

The design studio introduces the theme of hybrid urban typologies that draw from the space-making capacities of architecture, the service staging performance of infrastructure, and the territorial composition of landscape design to enhance the capacities of 21st century urban design policy and practice. The design course offers a range of critical design perspectives on the contemporary dynamics of world wide urbanization, with specific focus on the local impact of global urbanization processes on local built and lived spaces. By drawing on the distinct disciplinary know-how of architecture, civil engineering, and landscape design, urban design is reframed as something more than just the formal delivery of standardized spatial products for a mass constituency of consumers.

建筑理论课程
ARCHITECTURAL THEORY COURSES

本科二年级
建筑导论·赵辰 等
课程类型：必修
学时/学分：36学时/2学分

Undergraduate Program 2nd Year
INTRODUCTION GUIDE TO ARCHITECTURE • ZHAO Chen, etc.
Type: Required Course
Study Period and Credits:36 hours/2 credits

课程内容
1. 建筑学的基本定义
  第一讲：建筑与设计/赵辰
  第二讲：建筑与城市/丁沃沃
  第三讲：建筑与生活/张雷
2. 建筑的基本构成
  （1）建筑的物质构成
  第四讲：建筑的物质环境/赵辰
  第五讲：建筑与生态环境/吴蔚
  第六讲：建筑的材料观/王丹丹
  （2）建筑的文化构成
  第七讲：建筑与人文、艺术、审美/赵辰
  第八讲：建筑与环境景观/华晓宁
  第九讲：建筑的环境智慧/窦平平
  第十讲：建筑与身体经验/鲁安东
  （3）建筑师职业与建筑学术
  第十一讲：建筑与表现/赵辰
  第十二讲：建筑与几何形态/周凌
  第十三讲：建筑与数字技术/钟华颖
  第十四讲：建筑师的职业技能与社会责任/傅筱

Course Content
I Preliminary of architecture
Lecture 1: Architecture and design / ZHAO Chen
Lecture 2: Architecture and urbanization / DING Wowo
Lecture 3: Architecture and life / ZHANG Lei
II Basic attribute of architecture
II-1 Physical attribute
Lecture 4: Physical environment of architecture / ZHAO Chen
Lecture 5: Architecture and ecological environment / WU Wei
Lecture 6: Architecture with Materialization / Wang Dandan
II-2 Cultural attribute
Lecture 7: Architecture and civilization, arts, aesthetic / ZHAO Che
Lecture 8: Architecture and landscaping environment / HUA Xiaoning
Lecture 9: Environmental Intelligence in Architecture / DOU Pingpir
Lecture 10: Architecture and body / LU Andong
II-3 Architect: profession and academy
Lecture 11: Architecture and presentation / ZHAO Chen
Lecture 12: Architecture and geometrical form / ZHOU Ling
Lecture 13: Architectural and digital technology / ZHONG Huaying
Lecture 14: Architect's professional technique and responsibility / F
Xiao

本科三年级
建筑设计基础原理·周凌
课程类型：必修
学时/学分：36学时/2学分

Undergraduate Program 3rd Year
BASIC THEORY OF ARCHITECTURAL DESIGN
• ZHOU Ling
Type: Required Course
Study Period and Credits:36 hours/2 credits

教学目标
　　本课程是建筑学专业本科生的专业基础理论课程。本课程的任务主要是介绍建筑设计中形式与类型的基本原理。形式原理包含历史上各个时期的设计原则，类型原理讨论不同类型建筑的设计原理。
课程要求
　　1. 讲授大纲的重点内容；
　　2. 通过分析实例启迪学生的思维，加深学生对有关理论及其应用、工程实例等内容的理解；
　　3. 通过对实例的讨论，引导学生运用所学的专业理论知识，分析、解决实际问题。
课程内容
　　1. 形式与类型概述
　　2. 古典建筑形式语言
　　3. 现代建筑形式语言
　　4. 当代建筑形式语言
　　5. 类型设计
　　6. 材料与建造
　　7. 技术与规范
　　8. 课程总结

Training Objective
This course is a basic theory course for the undergraduate studen
of architecture. The main purpose of this course is to introduce th
basic principles of the form and type in architectural design. For
theory contains design principles in various periods of history; ty
theory discusses the design principles of different types of building
Course Requirement
1. Teach the key elements of the outline;
2. Enlighten students' thinking and enhance students' understandi
of the theories, its applications and project examples throu
analyzing examples;
3. Guide students using the professional knowledge to analysis a
solve practical problems through the discussion of examples.
Course Content
1. Overview of forms and types
2. Classical architecture form language
3. Modern architecture form language
4. Contemporary architecture form language
5. Type design
6. Materials and construction
7. Technology and specification
8. Course summary

本科三年级
居住建筑设计与居住区规划原理·冷天 刘铨
课程类型：必修
学时/学分：36学时/2学分

Undergraduate Program 3rd Year
**THEORY OF HOUSING DESIGN AND RESIDENTIAL
PLANNING** • LENG Tian, LIU Quan
Type: Required Course
Study Period and Credits:36 hours/2 credits

课程内容
　　第一讲：课程概述
　　第二讲：居住建筑的演变
　　第三讲：套型空间的设计
　　第四讲：套型空间的组合与单体设计（一）
　　第五讲：套型空间的组合与单体设计（二）
　　第六讲：居住建筑的结构、设备与施工
　　第七讲：专题讲座：住宅的适应性，支撑体住宅
　　第八讲：城市规划理论概述
　　第九讲：现代居住区规划的发展历程
　　第十讲：居住区的空间组织
　　第十一讲：居住区的道路交通系统规划与设计
　　第十二讲：居住区的绿地景观系统规划与设计
　　第十三讲：居住区公共设施规划、竖向设计与管线综合
　　第十四讲：专题讲座：住宅产品开发
　　第十五讲：专题讲座：住宅产品设计实践
　　第十六讲：课程总结，考试答疑

Course Content
Lecture 1: Introduction of the course
Lecture 2: Development of residential building
Lecture 3: Design of dwelling space
Lecture 4: Dwelling space arrangement and residential building desi
(1)
Lecture 5: Dwelling space arrangement and residential building desi
(2)
Lecture 6: Structure, detail, facility and construction of residential buildin
Lecture 7: Adapt ability of residential building, supporting house
Lecture 8: Introduction of the theories of urban planning
Lecture 9: History of modern residential planning
Lecture 10: Organization of residential space
Lecture 11: Traffic system planning and design of residential area
Lecture 12: Landscape planning and design of residential area
Lecture 13: Public facilities and infrastructure system
Lecture 14: Real estate development
Lecture 15: The practice of residential planning and housing desig
Lecture 16: Summary, question of the test

究生一年级
代建筑设计基础理论 · 张雷
程类型：必修
时/学分：18学时/1学分

Graduate Program 1st Year
PRELIMINARIES IN MODERN ARCHITECTURAL
DESIGN · ZHANG Lei
Type: Required Course
Study Period and Credits:18 hours/1 credit

教学目标
　　建筑可以被抽象到最基本的空间围合状态来面对它所必须解决的基本的适用问题，用最合理、最直接的空间组织和建造方式去解决问题，以普通材料和通用方法去回应复杂的使用要求，是建筑设计所应该关注的基本原则。
课程要求
　　1. 讲授大纲的重点内容；
　　2. 通过分析实例启迪学生的思维，加深学生对有关理论及其应用、工程实例等内容的理解；
　　3. 通过对实例的讨论，引导学生运用所学的专业理论知识，分析、解决实际问题。
课程内容
　　1. 基本建筑的思想
　　2. 基本空间的组织
　　3. 建筑类型的抽象与还原
　　4. 材料的运用与建造问题
　　5. 场所的形成及其意义
　　6. 建筑构思与设计概念

Training Objective
The architecture can be abstracted into spatial enclosure state to encounter basic application problems which must be settled. Solving problems with most reasonable and direct spatial organization and construction mode, and responding to operating requirements with common materials and general methods are basic principle concerned by building design.
Course Requirement
1. To teach key contents of syllabus;
2. To inspire students' thinking, deepen students' understanding on such contents as relevant theories and their application and engineering example through case analysis;
3. To guide students to use professional theories to analyze and solve practical problems through discussion of instances.
Course Content
1. Basic architectural thought
2. Basic spacial organization
3. Abstraction and restoration of architectural types
4. Utilization and construction of material
5. Formation of site and its meaning
6. Architectural conception and design concept

究生一年级
究方法与写作规范 · 鲁安东 胡恒 郜志
程类型：必修
时/学分：18学时/1学分

Graduate Program 1st Year
RESEARCH METHOD AND THESIS WRITING · LU
Andong, HU Heng, GAO Zhi
Type: Required Course
Study Period and Credits:18 hours/1 credit

教学目标
　　面向学术型硕士研究生的必修课程。它将向学生全面地介绍学术研究的特性、思维方式、常见方法以及开展学术研究必要的工作方式和写作规范。考虑到不同领域研究方法的差异，本课程的授课和作业将以专题的形式进行组织，包括建筑研究概论、设计研究、科学研究、历史理论研究4个模块。学生通过各模块的学习可以较为全面地了解建筑学科内主要的研究领域及相应的思维方式和研究方法。
课程要求
　　将介绍建筑学科的主要研究领域和当代研究前沿，介绍"研究"的特性、思维方式、主要任务、工作架构以及什么是好的研究，帮助学生建立对"研究"的基本认识；介绍文献检索和文献综述的规范和方法；介绍常见的定量研究、定性研究和设计研究的工作方法以及相应的写作规范。
课程内容
　　1. 综述
　　2. 文献
　　3. 科学研究及其方法
　　4. 科学研究及其写作规范
　　5. 历史理论研究及其方法
　　6. 历史理论研究及其写作规范
　　7. 设计研究及其方法
　　8. 城市规划理论概述

Training Objective
It is a compulsory course to MA. It comprehensively introduces features, ways of thinking and common methods of academic research, and necessary manner of working and writing standard for launching academic research to students. Considering differences of research methods among different fields, teaching and assignment of the course will be organized in the form of special topic, including four parts: introduction to architectural study, design study, scientific study and historical theory study. Through the study of all parts, students can comprehensively understand main research fields and corresponding ways of thinking and research methods of architecture.
Course Requirement
The course introduces main research fields and contemporary research frontier of architecture, features, ways of thinking and main tasks of "research", working structure of research, and definition of good research to help students form basic understanding of "research". The course also introduces standards and methods of literature retrieval and review, and working methods of common quantitative research, qualitative research and design research, and their corresponding writing standards.
Course Content
1. Review
2. Literature
3. Scientific research and methods
4. Scientific research and writing standards
5. Historical theory study and methods
6. Historical theory study and writing standards
7. Design research and methods
8. Overview of urban planning theory

城市理论课程
URBAN THEORY COURSES

本科四年级
城市设计及其理论 · 胡友培
课程类型：必修
学时/学分：36学时/2学分

Undergraduate Program 4th Year
URBAN DESIGN AND THEORY· HU Youpei
Type: Required Course
Study Period and Credits: 36 hours/2 credits

课程内容
第一讲：课程概述
第二讲：城市设计技术术语：城市规划相关术语；城市形态相关术语；城市交通相关术语；消防相关术语
第三讲：城市设计方法 —— 文本分析：城市设计上位规划；城市设计相关文献；文献分析方法
第四讲：城市设计方法 —— 数据分析：人口数据分析与配置；交通流量数据分析；功能分配数据分析；视线与高度数据分析；城市空间数据模型的建构
第五讲：城市设计方法 —— 城市肌理分类：城市肌理分类概述；肌理形态与建筑容量；肌理形态与开放空间；肌理形态与交通流量；城市绿地指标体系
第六讲：城市设计方法 —— 城市路网组织：城市道路结构与交通结构概述；城市路网与城市功能；城市路网与城市空间；城市路网与市政设施；城市道路断面设计
第七讲：城市设计方法 —— 城市设计表现：城市设计分析图；城市设计概念表达；城市设计成果解析图；城市设计地块深化设计表达；城市设计空间表达
第八讲：城市设计的历史与理论：城市设计的历史意义；城市设计理论的内涵
第九讲：城市路网形态：路网形态的类型和结构；路网形态与肌理；路网形态的变迁
第十讲：城市空间：城市空间的类型；城市空间结构；城市空间形态；城市空间形态的变迁
第十一讲：城市形态学：英国学派；意大利学派；法国学派；空间句法
第十二讲：城市形态的物理环境：城市形态与物理环境；城市形态与环境研究；城市形态与环境测评；城市形态与环境操作
第十三讲：景观都市主义：景观都市主义的理论、操作和范例
第十四讲：城市自组织现象及其研究：城市自组织现象的魅力与问题；城市自组织系统研究方法；典型自组织现象案例研究
第十五讲：建筑学图式理论与方法：图式理论的研究，建筑学图式的概念；图式理论的应用；作为设计工具的图式；当代城市语境中的建筑学图式理论探索
第十六讲：课程总结

Course Content
Lecture 1: Introduction
Lecture 2: Technical terms: terms of urban planning, urban morpholog urban traffic and fire protection
Lecture 3: Urban design methods — documents analysis: urba planning and policies; relative documents; document analys techniques and skills
Lecture 4: Urban design methods — data analysis: data analys of demography, traffic flow, function distribution, visual ar building height; modelling urban spatial data
Lecture 5: Urban design methods — classification of urba fabrics: introduction of urban fabrics; urban fabrics and floor are ratio; urban fabrics and open space; urban fabrics and traffic flo criteria system of urban green space
Lecture 6: Urban design methods — organization of urban roa network: introduction; urban road network and urban functio urban road network and urban space; urban road network ar civic facilities; design of urban road section
Lecture 7: Urban design methods — representation skills of urba Design: mapping and analysis; conceptual diagram; analytic representation of urban design; representation of detail desig spatial representation of urban design
Lecture 8: Brief history and theories of urban design: historic meaning of urban design; connotation of urban design theories
Lecture 9: Form of urban road network: typology, structure ar evolution of road network; road network and urban fabrics
Lecture 10: Urban space: typology, structure, morphology ar evolution of urban space
Lecture 11: Urban morphology: Cozen School; Italian Schoo French School; Space Syntax Theory
Lecture 12: Physical environment of urban forms: urban form and physical environment; environmental study; environmen evaluation and environmental operations
Lecture 13: Landscape urbanism: ideas, theories, operations ar examples of landscape urbanism
Lecture 14: Researches on the phenomena of the urban se organization: charms and problems of urban self-organizatic phenomena; research methodology on urban self-organizatic phenomena; case studies of urban self-organization phenomen
Lecture 15: Theory and method of architectural diagram: theoretic study on diagrams; concepts of architectural diagrams; applicatic of diagram theory; diagrams as design tools; theoretical resear of architectural diagrams in contemporary urban context
Lecture 16: Summary

本科四年级
景观规划设计及其理论 · 尹航
课程类型：选修
学时/学分：36学时/2学分

Undergraduate Program 4th Year
LANDSCAPE PALNNING DESIGN AND THEORY
· YIN Hang
Type: Elective Course
Study Period and Credits: 36 hours/2 credits

课程介绍
　　景观规划设计的对象包括所有的室外环境，景观与建筑的关系往往是紧密而互相影响的，这种关系在城市中表现得尤为明显。景观规划设计及理论课程希望从景观设计理念、场地设计技术和建筑周边环境塑造等方面开展课程的教学，为建筑学本科生建立更加全面的景观知识体系，并且完善建筑学本科生在建筑场地设计、总平面规划与城市设计等方面的设计能力。
　　本课程主要从三个方面展开：一是理念与历史：以历史的视角介绍景观学科的发展过程，让学生对景观学科有一个宏观的了解，初步理解景观设计理念的发展；二是场地与文脉：通过阐述景观规划设计与周边自然环境、地理位置、历史文脉和方案可持续性的关系，建立场地与文脉的设计思维；三是景观与建筑：通过设计方法授课、先例分析作业等方式让学生增强建筑的环境意识，了解建筑的场地设计的影响因素、一般步骤与设计方法，并通过与"建筑设计（六）"和"建筑设计（七）"的设计任务书相配合的同步课程设计训练来加强学生景观规划设计的能力。

Course Description
The object of landscape planning design includes all outdo environments; the relationship between landscape and buildi is often close and interactive, which is especially obvio in a city. This course expects to carry out teaching fro perspective of landscape design concept, site design technolo building's peripheral environment creation, etc., to establi a more comprehensive landscape knowledge system for t undergraduate students of architecture, and perfect their desi ability in building site design, master plane planning and urb design and so on.
This course includes three aspects:
1. Concept and history
2. Site and context
3. Landscape and building

研究生一年级
景观都市主义理论与方法 · 华晓宁
课程类型：选修
学时/学分：18学时/1学分

Graduate Program 1st Year
THEORY AND METHOD OF LANDSCAPE URBANISM
HUA Xiaoning
Type: Elective Course
Study Period and Credits: 18 hours / 1 credit

课程介绍

本课程作为国内首次以景观都市主义相关理论与策略为教学内容的尝试，介绍了景观都市主义思想产生的背景、缘起及其主要理论观点，并结合实例，重点分析了其在不同的场址和任务导向下发展起来的多样化的实践策略和操作性工具。

课程要求

1. 要求学生了解景观都市主义思想产生的背景、缘起和主要理念。
2. 要求学生能够初步运用景观都市主义的理念和方法分析和解决城市设计问题，从而在未来的城市设计实践中强化景观整合意识。

课程内容

第一讲：从图像到效能：景观都市实践的历史演进与当代视野
第二讲：生态效能导向的景观都市实践（一）
第三讲：生态效能导向的景观都市实践（二）
第四讲：社会效能导向的景观都市实践
第五讲：基础设施景观都市实践
第六讲：当代高密度城市中的地形学
第七讲：城市图绘与图解
第八讲：从原形到系统：AA景观都市主义

Course Description

Combining relevant theories and strategies of landscape urbanism firstly in China, the course introduces the background, origin and main theoretical viewpoint of landscape urbanism, and focuses on diversified practical strategies and operational tools developed under different orientations of site and task with examples.

Course Requirement

1. Students are required to understand the background, origin and main concept of landscape urbanism.
2. Students are required to preliminarily utilize the concept and method of landscape urbanism to analyze and solve the problem of urban design, so as to strengthen landscape integration consciousness in the future.

Course Content

Lecture 1: From pattern to performance: historical revolution and contemporary view of practice of landscape urbanism
Lecture 2: Eco-efficiency-oriented practice of landscape urbanism (1)
Lecture 3: Eco-efficiency-oriented practice of landscape urbanism (2)
Lecture 4: Social efficiency-oriented practice of landscape urbanism
Lecture 5: Infrastructure practice in landscape urbanism
Lecture 6: Geomorphology in contemporary high-density city
Lecture 7: Urban painting and diagrammatizing
Lecture 8: From prototype to system: AA landscape urbanism

研究生一年级
城市形态与建筑设计方法论 · 丁沃沃
课程类型：必修
学时/学分：36学时/2学分

Graduate Program 1st Year
URBAN FORM AND ARCHITECTURAL DESIGN METHODOLOGY · DING Wowo
Type: Required Course
Study Period and Credits: 36 hours / 2 credits

课程介绍

建筑学核心理论包括建筑学的认识论和设计方法论两大部分。建筑设计方法论主要探讨设计的认知规律、形式的逻辑、形式语言类型，以及人的行为、环境特征和建筑材料等客观规律对形式语言的选择及形式逻辑的构成策略。为此，设立了提升建筑设计方法的关于设计方法论的理论课程，作为建筑设计及其理论硕士学位的核心课程。

课程要求

1. 理解随着社会转型，城市建筑的基本概念在建筑学核心理论中的地位以及认知的视角。
2. 通过理论的研读和案例分析，理解建筑形式语言的成因和逻辑，并厘清中、西不同的发展脉络。
3. 通过研究案例的解析，理解建筑形式语言的操作并掌握设计研究的方法。

课程内容

第一讲：序言
第二讲：西方建筑学的基础
第三讲：中国：建筑的意义
第四讲：背景与文献研讨
第五讲：历史观与现代性
第六讲：现代城市形态演变与解析
第七讲：现代城市的"乌托邦"
第八讲：现代建筑的意义
第九讲：建筑形式的反思与探索
第十讲：建筑的量产与城市问题
第十一讲："乌托邦"的实践与反思
第十二讲：都市实践探索的理论价值
第十三讲：城市形态的研究
第十四讲：城市空间形态研究的方法
第十五讲：回归理性：建筑学方法论的新进展
第十六讲：建筑学与设计研究的意义
第十七讲：结语与研讨（一）
第十八讲：结语与研讨（二）

Course Description

The core theory of architecture includes epistemology and design methodology of Architecture. Architectural design methodology mainly discusses cognitive laws of design, logic of form and types of formal language, and the choice of formal language from objective laws such as human behavior, environmental feature and building material, and composition strategy of formal logic. Thus, the theory course about design methodology to promote architectural design methods is established as the core course of architectural design and theory master degree.

Course Requirement

1. To understand the status and cognitive perspective of basic concept of urban building in the core theory of architecture with the social transformation.
2. To understand the reason and logic of architectural formal language and different development process in China and West through reading theory and case analysis.
3. To understand the operation of architectural formal language and grasp methods of design study by analyzing study case.

Course Content

Lecture 1: Introduction
Lecture 2: Foundation of western architecture
Lecture 3: China: meaning of architecture
Lecture 4: Background and literature discussion
Lecture 5: Historicism and modernity
Lecture 6: Analysis and morphological evolution of modern city
Lecture 7: "Utopia" of modern city
Lecture 8: Meaning of modern Architecture
Lecture 9: Reflection and exploration of architectural form
Lecture 10: Mass production of buildings and urban problems
Lecture 11: Practice and reflection of "Utopia"
Lecture 12: Theoretical value of exploration on urban practice
Lecture 13: Study on urban morphology
Lecture 14: Method of urban spatial morphology study
Lecture 15: Return to rationality: new developments of methodology on architecture
Lecture 16: Meaning of architecture and design study
Lecture 17: Conclusion and discussion (1)
Lecture 18: Conclusion and discussion (2)

本科四年级
都市社会学 · 夏铸九
课程类型：必修
学时/学分：36学时/2学分

Undergraduate Program 4th Year
URBAN SOCIOLOGY · XIA Zhujiu
Type: Required Course
Study Period and Credits: 36 hours / 2 credits

课程介绍
　　第一部分以都市社会学的角度，对塑造空间结构的政策、规划以及设计的形成过程与结果，即都市与区域政策，包括美国、西欧及第三世界的不同的个案研究，提出比较分析。主题包括：区域空间结构与都市区域、对都市边缘性批判与都市非正式部门、住宅政策、都市运输、都市更新与保存、理论总结等。

课程内容
　　1.理论角度与方法论
　　2.1980年代美国的高科技与区域空间结构变迁
　　3.第二次世界大战后英国的新镇经验
　　4.都市化与依赖性：西非都市化的殖民模式
　　5.对都市边缘性批判与都市非正式部门：1970年代的墨西哥蒙特利市的违建聚落
　　6.1960年代与1970年代美国的邻里运动与1970年代西欧的都市运动
　　6.住宅政策：尼德兰的社会住宅经验
　　7.都市交通：1980年代委内瑞拉加拉卡斯的个案
　　8.都市的记忆与保存的里程碑：1970年意大利博洛尼亚的都市保存
　　9.理论总结（地域国家与全球化中国家的式微以及理论化规划）

Course Description
From the perspective of urban sociology, the first part present a comparative analysis of policy, planning of building space structure and forming process and result of design, that is, urb and regional policy, including research of different cases in U.S Western Europe and the third world. Subjects include: region spatial structure and urban area, criticism of urban marginal and urban informal sector, housing policy, urban transport, urb renewal and preservation, theoretical summary, etc.

Course Content
1. Theoretical perspective and methodology
2. High technology and regional spatial structure changes in t United States in the 1980s
3. New town experience in Britain after World War II
4. Urbanization and dependency: the colonial model urbanization in West Africa
5. Criticism of urban marginality and urban informal sector : illeg settlements in Monterey, Mexico in the 1970s
The neighborhood movement in the United States in the 196 and the urban movement in Western Europe in the 1970s
6. Housing policy: social housing experience in the Netherlands
7. Urban transport: the case of Caracas, Venezuela in the 1980
8. Urban memory and preservation milestone: urban preservati in Bologna, Italy in 1970
9. Theoretical summary (decline of regional countries a countries in globalization, and theoretical planning)

历史理论课程
HISTORY THEORY COURSES

本科二年级
中国建筑史（古代）· 赵辰 史文娟
课程类型：必修
学时/学分：36学时/2学分

Undergraduate Program 2nd Year
HISTORY OF CHINESE ARCHITECTURE (ANCIENT)
• ZHAO Chen
Type: Required Course
Study Period and Credits:36 hours / 2 credits

教学目标
　　本课程作为本科建筑学专业的历史与理论课程，目标在于培养学生的史学研究素养与对中国建筑及其历史的认识两个层面。在史学理论上，引导学生理解建筑史学这一交叉学科的多种楼面与视角，并从多种相关学科层面对学生进行基本史学研究方法的训练与指导。中国建筑史层面，培养学生对中国传统建筑的营造特征与文化背景建立构架性的认识体系。

课程内容
　　中国建筑史学七讲与方法论专题。七讲总体走向从微观到宏观，整体以建筑单体—建筑群体—聚落与城市—历史地理为序；从物质性到文化，建造技术—建造制度—建筑的日常性—纪念性—政治与宗教背景—美学追求。方法论专题包括建筑考古学、建筑技术史、人类学、美术史等层面。

Training Objective
As a mandatory historical & theoretical course for undergraduate students, this course aims at two aspects of training: the basic academic capability of historical research and the understanding of Chinses architectural history. It will help students to establish a knowledge frame, that the discipline of History of Architecture as a cross-discipline, is supported and enriched by multiple neighboring disciplines and that the features and development of Chinese Architecture roots deeply in the natural and cultural background.

Course Content
The course composes seven 4-hour lectures on Chinese Architecture and a series of lectures on methodology. The seven courses follow a route from individual to complex, from physical building to the intangible technique and to the cultural background, from technology to institution, to political and religious background and finally to aesthetic pursuit. The special topics on methodology include building archaeology, building science and technology, anthropology, art history and so on.

本科二年级
外国建筑史（古代）· 王丹丹
课程类型：必修
学时/学分：36学时/2学分

Undergraduate Program 2nd Year
HISTORY OF WORLD ARCHITECTURE (ANCIENT)
• WANG Dandan
Type: Required Course
Study Period and Credits: 36 hours / 2 credits

教学目标
　　本课程力图对西方建筑史的脉络做一个整体勾勒，使学生在掌握重要的建筑史知识点的同时，对西方建筑史在2000多年里的变迁的结构转折（不同风格的演变）有深入的理解。本课程希望引导学生对建筑史的发展与人类文明发展之间的密切关联有所认识。

课程内容
　　1. 概论　2. 希腊建筑　3. 罗马建筑　4. 中世纪建筑
　　5. 意大利的中世纪建筑　6. 文艺复兴　7. 巴洛克
　　8. 美国城市　9. 北欧浪漫主义　10. 加泰罗尼亚建筑
　　11. 先锋派　12. 德意志制造联盟与包豪斯
　　13. 苏维埃的建筑与城市　14. 1960年代的建筑
　　15. 1970年代的建筑　16. 答疑

Training Objective
This course seeks to give an overall outline of Western architectural history, so that the students may have an in depth understanding of the structural transition (different styles of evolution) of Western architectural history in the past 2000 years. This course hopes that students can understand the close association between the development of architectural history and the development of human civilization.

Course Content
1. Generality　2. Greek Architectures　3. Roman Architectures
4. The Middle Ages Architectures
5. The Middle Ages Architectures in Italy　6. Renaissance
7. Baroque　8. American Cities　9. Nordic Romanticism
10. Catalonian Architectures　11. Avant-Garde
12. German Manufacturing Alliance and Bauhaus
13. Soviet Architecture and Cities　14. 1960's Architectures
15. 1970's Architectures　16. Answer Questions

本科三年级
外国建筑史（当代）· 胡恒
课程类型：必修
学时/学分：36学时/2学分

Undergraduate Program 3rd Year
HISTORY OF WORLD ARCHITECTURE (MODERN) •
HU Heng
Type: Required Course
Study Period and Credits:36 hours / 2 credits

教学目标
　　本课程力图用专题的方式对文艺复兴时期的7位代表性的建筑师与5位现当代的重要建筑师作品做一细致的讲解。本课程将重要建筑师的全部作品尽可能在课程中梳理一遍，使学生能够全面掌握重要建筑师的设计思想、理论主旨、与时代的特殊关联、在建筑史中的意义。

课程内容
　　1. 伯鲁乃列斯基　2. 阿尔伯蒂　3. 伯拉孟特
　　4. 米开朗琪罗（1）　5. 米开朗琪罗（2）　6. 罗马诺
　　7. 桑索维诺　8. 帕拉蒂奥（1）　9. 帕拉蒂奥（2）
　　10. 赖特　11. 密斯　12. 勒·柯布西耶（1）
　　13. 勒·柯布西耶（2）　14. 海杜克　15. 妹岛和世
　　16. 答疑

Training Objective
This course seeks to make a detailed explanation to the works of 7 representative architects in the Renaissance period and 5 important modern and contemporary architects in a special way. This course will try to reorganize all works of these important architects, so that the students can fully grasp their design idea, theoretical subject and their particular relevance with the era and significance in the architectural history.

Course Content
1. Brunelleschi　2. Alberti　3. Bramante
4. Michelangelo(1)　5. Michelangelo(2)
6. Romano　7. Sansovino　8. Palladio(1)　9. Palladio(2)
10. Wright　11. Mies　12. Le Corbusier(1)　13. Le Corbusier(2)
14. Hejduk　15. Kazuyo Sejima
16. Answer Questions

本科三年级
中国建筑史（近现代）· 赵辰 冷天
课程类型：必修
学时/学分：36学时/2学分

Undergraduate Program 3rd Year
HISTORY OF CHINESE ARCHITECTURE (MODERN)
• ZHAO Chen, LENG Tian
Type: Required Course
Study Period and Credits:36 hours / 2 credits

课程介绍
　　本课程作为本科建筑学专业的历史与理论课程，是中国建筑史教学中的一部分。在中国与西方的古代建筑历史课程的基础上，了解中国社会进入近代，以至于现当代的发展进程。
　　在对比中西方建筑文化的基础之上，建立对中国近现代建筑的整体认识。深刻理解中国传统建筑文化在近代以来与西方建筑文化的冲突与相融之下，逐步演变发展至今天成为世界建筑文化的一部分之意义。

Course Description
As the history and theory course for undergraduate student of Architecture, this course is part of the teaching of History of Chinese Architecture. Based on the earlier studying of Chinese and Western history of ancient architecture, understand the evolution progress of Chinese society's entry into modern times and even contemporary age.
Based on the comparison of Chinese and Western building culture, establish the overall understanding of China's modern and contemporary buildings. Have further understanding the significance of China's traditional building culture's gradual evolution into one part of today's world building culture under conflict and blending with Western building culture in modern times.

研究生一年级
建筑理论研究 · 赵辰
课程类型：必修
学时/学分：18学时/1学分

Graduate Program 1st Year
STUDIES OF ARCHITECTURAL THEORY
ZHAO Chen
Type: Required Course
Study Period and Credits:18 hours / 1 credit

课程介绍
了解中、西方学者对中国建筑文化诠释的发展过程，理解新的建筑理论体系中对中国建筑文化重新诠释的必要性，学习重新诠释中国建筑文化的建筑观念与方法。

课程内容
1.本课的总览和基础
2.中国建筑：西方人的诠释与西方建筑观念的改变
3.中国建筑：中国人的诠释以及中国建筑学术体系的建立
4.木结构体系：中国建构文化的意义
5.住宅与园林：中国人居文化意义
6.宇宙观的和谐：中国城市文化的意义
7.讨论

Course Description
Understand the development process of Chinese and western scholars' interpretation of Chinese architectural culture, understand the necessity of reinterpretation of Chinese architectural culture in the new architectural theory system, and learn the architectural concepts and methods of reinterpretation of Chinese architectural culture.

Course Content
1. Overview and foundation of this course. 2. Chinese architecture: western interpretation and the change of western architectural concept. 3. Chinese architecture: Chinese interpretation and the establishment of Chinese architecture academic system. 4. Wood structure system: the significance of Chinese construction culture. 5. Residence and garden: the cultural significance of human settlement in China. 6. Harmony of cosmology: the significance of Chinese urban culture. 7. Discussion

---

研究生一年级
建筑理论研究 · 王骏阳
课程类型：必修
学时/学分：18学时/1学分

Graduate Program 1st Year
STUDIES OF ARCHITECTURAL THEORY · WANG Junyang
Type: Required Course
Study Period and Credits:18 hours / 1 credit

课程介绍
本课程是西方建筑史研究生教学的一部分。主要涉及当代西方建筑界具有代表性的思想和理论，其主题包括历史主义、先锋建筑、批判理论、建构文化以及对当代城市的解读等。本课程大量运用图片资料，广泛涉及哲学、历史、艺术等领域，力求在西方文化发展的背景中呈现建筑思想和理论的相对独立性及关联性，理解建筑作为一种人类活动所具有的社会和文化意义，启发学生的理论思维和批判精神。

课程内容
第一讲：建筑理论概论
第二讲：数字化建筑与传统建筑学的分离与融合
第三讲：语言、图解、空间内容
第四讲："拼贴城市"与城市的观念
第五讲：建构与营造
第六讲：手法主义与当代建筑
第七讲：从主线历史走向多元历史之后的思考
第八讲：讨论

Course Description
This course is a part of teaching Western architectural history for graduate students. It mainly deals with the representative thoughts and theories in Western architectural circles, including historicism, vanguard building, critical theory, tectonic culture and interpretation of contemporary cities etc.. Using a lot of pictures involving extensive fields including philosophy, history, art, etc., this course attempts to show the relative independence and relevance of architectural thoughts and theories under the development background of Western culture, understand the social and cultural significance owned by architectures as human activities, and inspire students' theoretical thinking and critical spirit.

Course Content
Lecture 1: Overview of architectural theories
Lecture 2: Separation and integration between digital architecture and traditional architecture
Lecture 3: Language, diagram and spatial content
Lecture 4: "Collage city" and concept of city
Lecture 5: Tectonics and yingzao (ying-tsao)
Lecture 6: Mannerism and modern architecture
Lecture 7: Thinking after main-line history to diverse history
Lecture 8: Discussion

研究生一年级
建筑史研究 · 胡恒
课程类型：选修
学时/学分：18学时/1学分

Graduate Program 1st Year
STUDIES IN ARCHITECTURAL HISTORY · HU Heng
Type: Elective Course
Study Period and Credits: 18 hours / 1 credit

教学目标
促进学生对历史研究的主题、方法、路径有初步的认识，通过具体的案例讲解使学生能够理解当代中国建筑史研究的诸多可能性。

课程内容
1.图像与建筑史研究（1-文学、装置、设计）
2.图像与建筑史研究（2-文学、装置、设计）
3.图像与建筑史研究（3-绘画与园林）
4.图像与建筑史研究（4-绘画、建筑、历史）
5.图像与建筑史研究（5-文学与空间转译）
6.方法讨论1
7.方法讨论2

Training Objective
Promote students' preliminary understanding of the topic, method and approach of historical research. Make students understand the possibilities of contemporary study on history of Chinese architecture through specific cases.

Course Content
1. Image and architectural history study (1-literature, device and design)
2. Image and architectural history study (2-literature, device and design)
3. Image and architectural history study (3-painting and garden)
4. Image and architectural history study (4-painting, architecture and history)
5. Image and architectural history study (5-literature and spatial transform)
6. Method discussion 1
7. Method discussion 2

研究生一年级
中国建构文化研究 · 赵辰
课程类型：选修
学时/学分：18学时/1学分

Graduate Program 1st Year
STUDIES IN CHINESE WOODEN TECTONIC CULTURE · ZHAO Chen
Type: Elective Course
Study Period and Credits: 18 hours / 1 credit

教学目标
以木为材料的建构文化是世界各文明中的基本成分，中国的木建构文化更是深厚而丰富。在全球可持续发展要求之下，木建构文化必须得到重新的认识和评价。对于中国建筑文化来说，更具有文化传统再认识和再发展的意义。

课程内容
阶段一：理论基础：对全球木建构文化的重新认识
阶段二：中国木建构文化的原则和方法（讲座与工作室）
阶段三：中国木建构的基本形：从家具到建筑（讲座与工作室）
阶段四：结构造型的发展和木建构的现代化（讲座）
阶段五：建造实验的鼓动（讲座与工作室）

Training Objective
The wood - based construction culture is the basic component of all civilizations in the world, and Chinese wood construction culture is profound and abundant. Under the requirement of global sustainable development, wood construction culture must be re-recognized and evaluated. For Chinese architectural culture, it is of great significance to re-recognize and re-develop the cultural tradition.

Course Content
Stage 1: theoretical basis: re-understanding of global wood construction culture
Stage 2: principles and methods of Chinese wood construction culture (lectures and studios)
Stage 3: the basic shape of Chinese wood construction: from furniture to architecture (lectures and studios)
Stage 4: development of structural modeling and modernization of wood construction (lectures)
Stage 5: agitation of construction experiment (lectures and studios)

建筑技术课程
ARCHITECTURAL TECHNOLOGY COURSES

本科二年级
CAAD理论与实践（一）·童滋雨
课程类型：必修
学时/学分：36学时/2学分

Undergraduate Program 2nd Year
THEORY AND PRACTICE OF CAAD 1 • TONG Ziyu
Type: Required Course
Study Period and Credits: 36 hours / 2 credits

课程介绍
　　在现阶段的CAD教学中，强调了建筑设计在建筑学教学中的主干地位，将计算机技术定位于绘图工具，本课程就是帮助学生可以尽快并且熟练地掌握如何利用计算机工具进行建筑设计的表达。课程中整合了CAD知识、建筑制图知识以及建筑表现知识，将传统CAD教学中教会学生用计算机绘图的模式向教会学生用计算机绘制有形式感的建筑图的模式转变，强调准确性和表现力作为评价CAD学习的两个最重要指标。
　　本课程的具体学习内容包括：
　　1. 初步掌握AutoCAD软件和SketchUP软件的使用，能够熟练完成二维制图和三维建模的操作；
　　2. 掌握建筑制图的相关知识，包括建筑投影的基本概念、平立剖面、轴测、透视和阴影的制图方法和技巧；
　　3. 图面效果表达的技巧，包括黑白线条图和彩色图纸的表达方法和排版方法。

Course Description
The core position of architectural design is emphasized in the CAD course. The computer technology is defined as drawing instrument. The course helps students learn how to make architectural presentation using computer fast and expertly. The knowledge of CAD, architectural drawing and architectural presentation are integrated into the course. The traditional mode of teaching student to draw in CAD course will be transformed into teaching student to draw architectural drawing with sense of form. The precision and expression will be emphasized as two most important factors to estimate the teaching effect of CAD course.
Contents of the course include:
1. Use AutoCAD and SketchUP to achieve the 2-D drawing and 3-D modeling expertly.
2. Learn relational knowledge of architectural drawing, including basic concepts of architectural projection, drawing methods and skills of plan, elevation, section, axonometry, perspective and shadow.
3. Skills of presentation, including the methods of expression and lay out using mono and colorful drawings

---

本科三年级
建筑技术（一）：结构、构造与施工·傅筱
课程类型：必修
学时/学分：36学时/2学分

Undergraduate Program 3rd Year
ARCHITECTURAL TECHNOLOGY 1: STRUCTURE, CONSTRUCTION AND EXECUTION • FU Xiao
Type: Required Course
Study Period and Credits: 36 hours / 2 credits

课程介绍
　　本课程是建筑学专业本科生的专业主干课程。本课程的任务主要是以建筑师的工作性质为基础，讨论一个建筑生成过程中最基本的三大技术支撑（结构、构造、施工）的原理性知识要点，以及它们在建筑实践中的相互关系。

Course Description
The course is a major course for the undergraduate students architecture. The main purpose of this course is based on the nature of the architect's work, to discuss the principle knowledge points of the basic three technical supports in the process of generating construction (structure, construction, execution), and their mutual relations in the architectural practice.

本科三年级
建筑技术（二）：建筑物理·吴蔚
课程类型：必修
学时/学分：36学时/2学分

Undergraduate Program 3rd Year
ARCHITECTURAL TECHNOLOGY 2: BUILDING PHYSICS • WU Wei
Type: Required Course
Study Period and Credits: 36 hours / 2 credits

课程介绍
　　本课程是针对三年级学生所设计，课程介绍了建筑热工学、建筑光学、建筑声学中的基本概念和基本原理，使学生能掌握建筑的热环境、声环境、光环境的基本评估方法，以及相关的国家标准。完成学业后在此方向上能阅读相关书籍，具备在数字技术方法等相关资料的帮助下，完成一定的建筑节能设计的能力。

Course Description
Designed for the Grade 3 students, this course introduces the basic concepts and basic principles in architectural thermal engineering architectural optics and architectural acoustics, so that the student can master the basic methods for the assessment of building thermal environment, sound environment and light environment as well as the related national standards. After graduation, the student will be able to read the related books regarding these aspects and have the ability to complete certain building energy efficient designs with the help of the related digital techniques and methods

---

本科三年级
建筑技术（三）：建筑设备·吴蔚
课程类型：必修
学时/学分：36学时/2学分

Undergraduate 3rd Year
ARCHITECTURAL TECHNOLOGY 3: BUILDING EQUIPMENT • WU Wei
Type: Required Course
Study Period and Credits: 36 hours / 2 credits

课程介绍
　　本课程是针对南京大学建筑与城市规划学院本科三年级学生所设计。课程介绍了建筑给水排水系统、采暖通风与空气调节系统、电气工程的基本理论、基本知识和基本技能，使学生能熟练地阅读水电、暖通工程图，熟悉水电及消防的设计、施工规范，了解燃气供应、安全用电及建筑防火、防雷的初步知识。

Course Description
This course is an undergraduate class offered in the School of Architecture and Urban Planning, Nanjing University. It introduces the basic principle of the building services systems, the technique of integration amongst the building services and the building. Throughout the course, the fundamental importance to energy, ventilation, air-conditioning and comfort in buildings are highlighted

---

研究生一年级
传热学与计算流体力学基础·郜志
课程类型：选修
学时/学分：18学时/1学分

Graduate Program 1st Year
FUNDAMENTALS OF HEAT TRANSFER AND COMPUTATIONAL FLUID DYNAMICS • GAO Zhi
Type: Elective Course
Study Period and Credits: 18 hours / 1 credit

课程介绍
　　本课程的主要任务是使建筑学/建筑技术学专业的学生掌握传热学和计算流体力学的基本概念和基础知识，通过课程教学，使学生熟悉传热学中导热、对流和辐射的经典理论，了解传热学和计算流体力学的实际应用和最新研究进展，为建筑能源和环境系统的计算和模拟打下坚实的理论基础。教学中尽量简化传热学和计算流体力学经典课程中复杂公式的推导过程，而着重于如何解决建筑能源与建筑环境中涉及流体流动和传热的实际应用问题。

Course Description
This course introduces students majoring in building science and engineering / building technology to the fundamentals of heat transfer and computational fluid dynamics (CFD). Students will study classical theories of conduction, convection and radiation heat transfers, and learn advanced research developments of heat transfer and CFD. The complex mathematics and physical equations are not emphasized. It is desirable that for real-case scenarios students will have the ability to analyze flow and heat transfer phenomena in building energy and environment systems

---

研究生一年级
GIS基础与应用·童滋雨
课程类型：必修
学时/学分：18学时/1学分

Graduate Program 1st Year
CONCEPT AND APPLICATION OF GIS • TONG Ziyu
Type: Required Course
Study Period and Credits:18 hours / 1 credit

课程介绍
　　本课程的主要目的是让学生理解GIS的相关概念以及GIS对城市研究的意义，并能够利用GIS软件对城市进行分析和研究。

Course Description
This course aims to enable students to understand the relational concepts of GIS and the significance of GIS to urban research and to be able to use GIS software to carry out urban analysis and research.

研究生一年级
建筑环境学 · 郜志
课程类型: 选修
学时/学分: 18学时/1学分

Graduate Program 1st Year
ARCHITECTURAL ENVIRONMENTAL SCIENCE · GAO Zhi
Type: Elective Course
Study Period and Credits:18 hours / 1 credit

课程介绍

本课程的主要任务是使建筑学/建筑技术学专业的学生掌握建筑环境的基本概念，学习建筑与城市热湿环境、风环境和空气质量的基础知识。通过课程教学，使学生熟悉城市微气候等理论，并了解人体对热湿环境的反应，掌握建筑环境学的实际应用和最新研究进展，为建筑能源和环境系统的测量与模拟打下坚实的基础。

Course Description

This course introduces students majoring in building science and engineering / building technology to the fundamentals of built environment. Students will study classical theories of built / urban thermal and humid environment, wind environment and air quality. Students will also familiarize urban micro environment and human reactions to thermal and humid environment. It is desirable that students will have the ability to measure and simulate building energy and environment systems based upon the knowledge of the latest development of the study of built environment.

研究生一年级
材料与建造 · 冯金龙
课程类型: 必修
学时/学分: 18学时/1学分

Graduate Program 1st Year
MATERIALS AND CONSTRUCTION · FENG Jinlong
Type: Required Course
Study Period and Credits:18 hours / 1 credit

课程介绍

本课程将介绍现代建筑技术的发展过程，论述现代建筑技术在建筑设计中的重要作用；探讨由材料、结构和构造方式所形成的建筑建造的逻辑方式；研究建筑形式产生的物质技术基础，诠释现代建筑的建构理论与研究方法。

Course Description

It introduces the development process of modern architecture technology and discusses the important role played by the modern architecture technology and its aesthetic concepts in the architectural design. It explores the logical methods of construction of the architecture formed by materials, structure and construction. It studies the material and technical basis for the creation of architectural form, and interprets construction theory and research methods for modern architectures.

研究生一年级
计算机辅助建筑设计技术 · 吉国华
课程类型: 必修
学时/学分: 36学时/2学分

Graduate Required 1st Year
TECHNOLOGY OF CAAD · JI Guohua
Type: Required Course
Study Period and Credits:36 hours / 2 credits

课程介绍

随着计算机辅助建筑设计技术的快速发展，当前数字技术在建筑设计中的角色逐渐从辅助绘图转向了真正的辅助设计，并引发了设计的革命和建筑的形式创新。本课程讲授Grasshopper参数化编程建模方法以及相关的几何知识，让学生在掌握参数化编程建模技术的同时，增强以理性的过程思维方式分析和解决设计问题的能力，为数字建筑设计和数字建造打下必要的基础。

基于Rhinoceros的算法编程平台Grasshopper的参数化建模方法，讲授内容包括各类运算器的功能与使用、图形的生成与分析、数据的结构与组织、各类建模的思路与方法，以及相应的数学与计算机编程知识。

Course Description

The course introduces methods of Grasshopper parametric programming and modeling and relevant geometric knowledge. The course allows students to master these methods, and enhance the ability to analyze and solve design problems with rational thinking at the same time, building necessary foundation for digital architecture design and digital construction.

In this course, the teacher will teach parametric modeling methods based on Grasshopper, a algorithmic programming platform for Rhinoceros, including functions and application of all kinds of arithmetic units, pattern formation and analysis, structure and organization of data, various thoughts and methods of modeling, and related knowledge of mathematics and computer programming.

研究生一年级
建筑体系整合 · 吴蔚
课程类型: 选修
学时/学分: 18~36学时/1~2学分

Graduate Program 1st Year
ADVANCED BUILDING SYSYTEM INTEGRATION · WU Wei
Type: Elective Course
Study Period and Credits: 18~36 hours / 1~2 credits

课程介绍

本课程是从建筑各个体系整合的角度来解析建筑设计。首先，课程介绍了建筑体系整合的基本概念、原理及其美学观念；然后具体解读以上各个设计元素在整个建筑体系中所扮演的角色及其影响力；最后，以全球的环境问题和人类生存与发展为着眼点，引导同学们重新审视和评判我们奉为信条的设计理念和价值系统。本课程着重强调建筑设计需要了解不同学科和领域的知识，熟悉各工种之间的配合和协调。

Course Description

A building is an assemblage of materials and components to obtain a shelter from external environment with a certain amount of safety so as to provide a suitable internal environment for physiological and psychological comfort in an economical manner. This course examines the role of building technology in architectural design, shows how environmental concerns have shaped the nature of buildings, and takes a holistic view to understand the integration of different building systems. It employs total building performance which is a systematic approach, to evaluate the performance of various sub-systems and to appraise the degree of integration of the sub-systems.

研究生一年级
算法设计 · 吉国华
课程类型: 选修
学时/学分: 18~36学时/1~2学分

Graduate Program 1st Year
ALGORITHMIC DESIGN· JI Guohuai
Type: Elective Course
Study Period and Credits: 18~36 hours / 1~2 credits

课程介绍

编程技术是数字建筑的基础，本课程主要讲授Grasshopper脚本编程和处理编程，让学生在掌握代码编程基础技术的同时，增强以理性的过程思维方式分析和解决问题的能力，逐步掌握数字设计的方法，为数字设计和建造课程打好基础。

Course Description

Programming technology is the foundation of digital architecture, this course mainly teaches Grasshopper script programming and processing programming, so that students can master the basic technology of code programming, at the same time, enhance the ability to analyze and solve design problems with rational thinking at the same time, gradually master the method of digital design, and lay a good foundation for the course of digital design and construction.

研究生二年级
建设工程项目管理 · 谢明瑞
课程类型：选修
学时/学分：36学时/2学分

Graduate Program 2nd Year
MANAGEMENT OF CONSTRUCTION PROJECT •XIE Mingrui
Type: Elective Course
Study Period and Credits: 36 hours / 2 credits

课程介绍
　　帮助学生系统掌握建设工程项目管理的基本概念、理论体系和管理方法，了解建筑规划设计在建设工程项目中的地位、特点和重要性。
　　延展建筑学专业学生基本知识结构层面，拓展学生的发展方向。

Course Description
To help students systematically master the basic concept, theoretica system and management method of construction engineerin project management, understand the position, characteristics an importance of architectural planning design in the constructio engineering project.
To extend the basic knowledge structure level of students majorin in architecture, develop the development direction of students.

研究生二年级
建筑环境学与设计 · 尤伟 郜志
课程类型：必修
学时/学分：36学时/2学分

Graduate Program 2nd Year
ARCHITECTURAL ENVIRONMENTAL SCIENCE AND DESIGN• YOU Wei, GAO Zhi
Type: Required Course
Study Period and Credits: 36 hours / 2 credits

课程介绍
　　本课程是基于建筑环境学课程的设计实践课程，旨在将建筑环境学课程的理论知识通过设计案例的练习加以运用，加深对建筑环境学知识的理解，并训练如何通过设计优化营造良好的室内环境品质。
　　课程分为授课和案例设计练习两部分。授课部分介绍目前关于被动式设计研究成果、工程实践案例中的被动式设计方法以及软件模拟分析技术；案例设计练习教授学生学习基于性能评估的优化设计方法，选取学生较为熟悉的住宅、幼儿园等体量较小的建筑类型作为设计优化对象，通过软件分析发现现有的室内环境设计的不足，并基于现有研究成果知识提出优化策略，最后通过软件模拟加以验证。课程要求学生将建筑环境学课程所学知识用于本设计课程的室内环境品质的量化及控制。本课程着重训练建筑设计与环境工程学科知识的配合。

Course Description
This course is a design practice course based on the course c building environment, aiming to apply the theoretical knowledg of the building environment course through the practice of desig cases, so as to deepen the understanding of the knowledg of building environment, and train how to create a good indoc environment quality through design optimization.
The course is divided into two parts: teaching and case desig practice. The teaching part introduces current research results c passive design, passive design methods in engineering practic cases and software simulation analysis technology. The cas design practice part teaches students to study optimal desig method based on performance evaluation, choose the residenc kindergarten and other building with small volume that studen are more familiar with as the design optimization objects, fin existing deficiency in indoor environment design through softwa analysis, propose optimization strategy according to existin research result, and finally verify through software simulatio The course requires students to apply what they have learned the building environment course to the quantification and contr of indoor environment quality in this design course.This cours focuses on the integration of architectural design and environment engineering knowledge.

研究生三年级
建筑学中的技术人文主义 · 窦平平
课程类型：必修
学时/学分：36学时/2学分

Graduate Program 3rd Year
TECHNOLOGY OF HUMANISM IN ARCHITECTURE • DOU Pingping
Type: Required Course
Study Period and Credits: 36 hours / 2 credits

课程介绍
　　课程详尽地阐释了为满足建筑的多方需求而投入的技术探索和人文关怀。课程包括四大版块，共十六个主题讲座，以案例精读的形式引介相关建筑师和学者的作品和理论。希望培养学生对建筑学中的技术议题进行批判性和人文主义的深入理解。

Course Description
This course elaborates the technological endeavors and humanis concern in fulfilling the multifaceted architectural demands. It tak shape in a series of sixteen themed lectures, grouped in fo sections, introducing prominent architects and scholars throug richly illustrated case studies and interpretations. It aims to nurtu the students with a critical and humanistic understanding of the r of technology playing in the discipline of architecture.

其他
MISCELLANEA

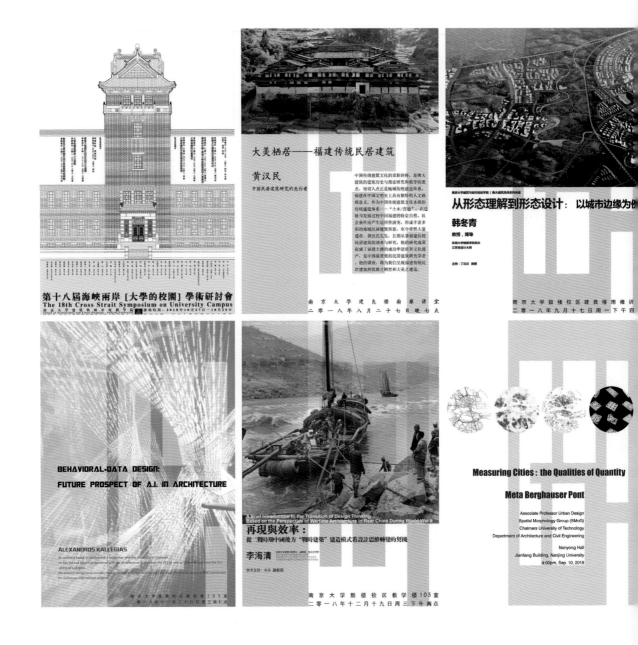

第十八届海峡两岸 [大学的校園] 學術研討會
The 18th Cross Strait Symposium on University Campus
南京大学建筑与城市规划学院

大美栖居——福建传统民居建筑

黄汉民
中国民居建筑研究的先行者

南京大学建良楼南座讲堂
二零一八年八月二十七日晚七点

从形态理解到形态设计：以城市边缘为例

韩冬青
教授，博导
东南大学建筑学院院长
江苏省设计大师
主持：丁沃沃 教授

南京大学鼓楼校区建良楼南座讲堂
二零一八年九月十七日周一下午四

BEHAVIORAL-DATA DESIGN:
FUTURE PROSPECT OF A.I. IN ARCHITECTURE

ALEXANDROS KALLEGIAS

南京大学鼓楼校区城市规划205室
二零一八年十二月二十八日周三晚七点

A Brief Introduction to the Transition of Design Thinking
Based on the Perspective of Wartime Architecture in Rear China During World War II

再現與效率：
從二戰時期中國後方 "戰時建築" 建造模式看設計思維轉變的契機

李海清
学术主持 冷天 副教授

南京大学鼓楼校区教学楼105室
二零一八年十二月十九日周三下午两点

Measuring Cities : the Qualities of Quantity

Meta Berghauser Pont

Associate Professor Urban Design
Spatial Morphology Group (SMoG)
Chalmers University of Technology
Department of Architecture and Civil Engineering

Nanyong Hall
Jianliang Building, Nanjing University
4:00pm, Sep. 10, 2018

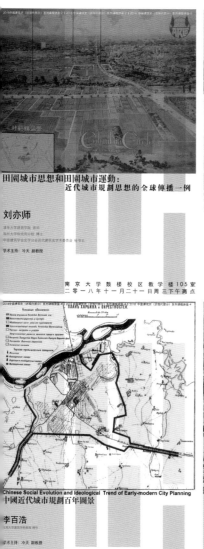

田園城市思想和田園城市運動：
近代城市規劃思想的全球傳播一例

刘亦师

南京大学建筑学院博士后
加拿大麦吉尔大学访问博士
中国建筑历史学会近代建筑史学术委员会会员等

学术主持：冷天 副教授

南 京 大 学 鼓 楼 校 区 教 学 楼 105 室
二 零 一 八 年 十 一 月 二 十 一 日 周 三 下 午 两 点

南京大学建筑与城市规划学院 ｜ 南大建筑高峰系列讲座

现代木结构研究与应用

刘伟庆

刘工：教授、博士生导师。南京工业大学副校长，江苏省土木建筑学会理事等，土木建筑学科首席教授，江苏省土木工程与防灾减灾重点实验室主任，江苏省绿色建筑中心主任，江苏省绿色高性能材料与结构工程实验室主任。

主要从事此木结构、复合材料结构、结构减震控制等研究。主持国家基金联合重点项目绿色生态木竹结构体系研究与示范应用、973 课题、动态振动环境中的木下结构材料体系革命原性演化机理、国家自然科学基金重点项目 - 新型村镇钢铁结构复合材料体系新创研与关键基础理论科研课题 20 余项，发表学术论文 500 余篇，其中 SCI、EI 收录论文 200 余篇，主编规程 5 部。授权国际专利及国内专利 40 余件，采用新材料获二等奖 1 项，江苏省科技进步一等奖 4 项。省部级科技进步二等奖 2 项，全国优秀建筑结构设计一等奖 3 项，江苏省"333高层次人才第二工程"中青年科技领军人才。

主持：丁沃沃教授

南 京 大 学 鼓 楼 校 区 建 良 楼 南 雍 讲 堂
二 零 一 八 年 十 二 月 三 日 星 期 一 下 午 四 点

南京大学建筑与城市规划学院 ｜ 南大建筑高峰系列讲座

The way of architecture

孔宇航

教授，博导

天津大学建筑学院党委副书记

主持：丁沃沃 教授

南 京 大 学 鼓 楼 校 区 建 良 楼 南 雍 讲 堂
二 零 一 八 年 十 月 十 五 日 下 午 六 点 半 至 八 点

Chinese Social Evolution and Ideological Trend of Early-modern City Planning

中國近代城市規劃百年圖景

李百浩

同济大学建筑学院教授 博导

学术主持：冷天 副教授

南 京 大 学 鼓 楼 校 区 教 学 楼 105 室
二 零 一 八 年 十 二 月 十 二 日 周 三 下 午 两 点

城市化進程中的上海近代建築

卢永毅

同济大学建筑与城市规划学院 教授 博导

学术主持：冷天 副教授

南 京 大 学 鼓 楼 校 区 教 学 楼 105 室
二 零 一 八 年 十 二 月 五 日 周 三 下 午 两 点

Design Optimization in Architecture

建築最適化

Hyoung-June Park

Associate Professor
School of Architecture, University of Hawaii at Manoa
Director of Design Future[s] Lab

夏威夷大学（马诺亚校区）建筑学院副教授
未来设计实验室主任

2019年5月24日 19:00
南京大学建良楼南雍讲堂

南京大学建筑与城市规划学院
School of Architecture and Urban Planning Nanjing University

硕士学位论文列表
List of Thesis for Master Degree

研究生姓名	研究生论文标题	导师姓名
梁庆华	南京三江大学学生宿舍设计及公共空间研究	张 雷
裘嘉珺	沙头村乡村书屋设计及木材和茅草在当代乡土实践中的研究	张 雷
王姝宁	福州登云艺术中心设计及其三维曲面幕墙建构研究	张 雷
于 昕	南京华侨城OCAT艺术中心设计及其公共空间和业态布局研究	张 雷
袁子燕	深汕合作区赤石公园一期商业街设计及模块化钢结构建筑研究	张 雷
章 程	腾冲玛御谷多功能文化中心设计及地域性材料和公共空间研究	冯金龙
刘 宣	南京理工大学江阴校区图书馆建筑设计	冯金龙
王浩哲	南京大学仙林校区众创空间二期建筑设计	冯金龙
于明霞	南京理工大学致知楼适应性再利用设计	冯金龙
臧 倩	南京市利济巷城市公共文化空间装配式建筑设计	吉国华
黄陈瑶	基于优化算法的产业园自动布局研究	吉国华
谢灵晋	基于刚性折叠的可展开面造型生成方法与工具研究	吉国华
张馨元	基于多智能体的建筑群体自动布局研究——以产业园为例	吉国华
朱鼎祥	曲面三角形网格单元划分的均匀化研究	吉国华
陈 妍	3D混凝土打印在建筑设计中的应用——以某传达室项目为例	傅 筱
刘江全	基于机械臂热线切割EPS的造型研究	傅 筱
李潇乐	游客中心公共空间精细化设计研究——茅山游客中心建筑室内设计	傅 筱
孙鸿鹏	干作业改造既有建筑的可能性研究——以福建长汀南雍九屋改造设计为例	傅 筱
徐雅甜	基于教学角度的中大跨建筑设计研究——大学生健身中心改扩建设计	傅 筱
袁 一	生物医药企业加速器建筑设计研究——南京生物医药谷聚智园项目建筑设计	周 凌
赵霏霏	历史街区容积率与空间品质关联性研究——福建长汀水东街地块1建筑设计	周 凌
代晓荣	淄博青城县历史空间复原研究及设计	周 凌
杨瑞东	宁镇地区乡村建筑建造体系及其更新策略研究——以南京徐家院村为例	周 凌
陈思涵	源于传统聚落组织形态的低层高密度居住建筑集群设计方法研究	周 凌
李恬楚	元阳哈尼梯田遗产区传统村落改造规划设计及修复性导则	丁沃沃
刘姿佑	基于院落风环境的多层围合式建筑设计策略研究	丁沃沃
赵媛倩	经济发达地区基于日照分析的多层住宅建筑形态优化研究——以南京市为例	丁沃沃
从 彬	基于三维视球数据分析的街道空间量化研究	丁沃沃
耿蒙蒙	基于街景图的城市街道空间测度方法研究	丁沃沃
马亚菲		
熊 攀		
徐新杉		

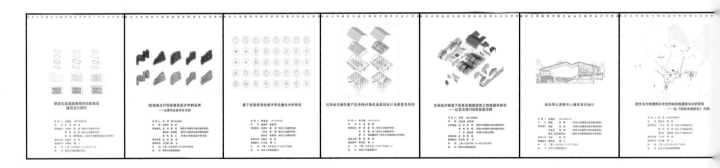

研究生姓名	研究生论文标题	导师姓名
吴 帆	石臼湖—固城湖圩区景观格局与聚落形态研究	华晓宁
王一侬	乡村振兴中新型公共交往空间设计研究——安徽省郎溪县定埠镇原服装厂节点改造更新设计	华晓宁
吴家禾	基于民俗事件的乡镇空间活化——安徽省郎溪县定埠镇公共空间系统改造更新设计	华晓宁
陈欣冉	近代南京沿街商业建筑立面的建造方式研究——以"三十四标"项目为例	赵 辰
曹永青	近现代优秀建筑再利用引发的城市更新——以南京庆和昌记地块为例	赵 辰
刘 刚	工业遗产与历史城市地段的空间形态整合——以南京大报恩寺遗址公园与1865创意产业园衔接地带为例	赵 辰
王丽丽	作为非标准化住宿的中国民宿研究	王骏阳
桂 喻	图文互文地理图示中空间知识的提取与分析研究——以《洪武京城图志》为例	鲁安东
梅凯强	日本竞进社模范蚕室对民国时期闽粤地区改良蚕室的影响及其相互比较	鲁安东
苏 彤	基于场所历史研究的南京长江大桥南堡公园更新设计	鲁安东
宫传佳	基于街景照片与图像识别的街道日照时数研究	童滋雨
王 坦	基于局地气候分区的宏观尺度城市形态建模及城市微气候数值模拟方法研究	童滋雨
董晶晶	基于剖面视角的城市形态量化分析研究	童滋雨
徐 沙	基于大数据与城市复杂理论工具的公园绿地布局评价模型研究	童滋雨
程 睿	米开朗基罗的军事建筑研究	胡 恒
金璐璐	《罗马古代遗迹》与《教堂遗迹介绍》——从帕拉第奥的导游手册看文艺复兴建筑师对古罗马城市的认知	胡 恒
文 涵	从文学到建筑——以李渔《十二楼》中的叙事、结构为例	胡 恒
聂柏慧	基于GH平台的天然光非视觉效应评价工具研究	吴 蔚
汤 晋	天然采光模拟插件在复杂采光模拟空间的适用性研究	吴 蔚
王照宇	室内多孔材料对环境湿度及建筑能耗的影响——以羊毛保温材料为例	秦孟昊
芮丽燕	城市街道交叉口的形态及绿化对通风的影响	郜 志
杨 喆	南京地区住宅室内臭氧的影响、来源与贡献	郜 志
陈 硕	公众参与下的工业遗产保护性再利用——以谢菲尔德波特兰工场为例	窦平平 丁沃沃
李鹏程	简洁形式后的设计逻辑及其建造实现——以江苏省产业技术研究院建设项目为例	冯金龙 谢明瑞
王 永	BIM技术在江苏省产业技术研究院建设项目中的应用研究	冯金龙 谢明瑞

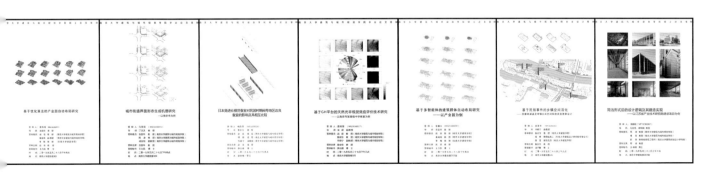

本科生 Undergraduate

2015级学生
Students 2015

卞秋怡 BIAN Qiuyi	陈景杨 Chen Jingyang	刘 越 LIU Yue	仙海斌 XIAN Haibin
吕文倩 LV Wenqian	邸晓宇 DI Xiaoyu	罗逍遥 LUO Xiaoyao	杨鑫毓 YANG Xinyu
龚 正 GONG Zheng	丁展图 DING Zhantu	秦伟航 QIN Weihang	杨 洋 YANG Yang
顾卓琳 GU Zhuolin	顾梦婕 GU Mengjie	沈静雯 SHEN Jingwen	叶庆锋 YE Qingfeng
戴添趣 DAI Tianqu	何 璇 HE Xuan	汪 榕 WANG Rong	张昊阳 ZHANG Haoyang
赵 彤 ZHAO Tong	李博文 LI Bowen	王雪梅 WANG Xuemei	周 杰 ZHOU Jie
罗紫璇 LUO Zixuan	李心仪 LI Xinyi	卫 斌 WEI Bin	

2016级学生
Students 2016

陈 帆 CHEN Fan	黄靖绮 HUANG Jingqi	丘雨辰 QIU Yuchen	余沁蔓 YU Qinman
陈婧秋 CHEN Jingqiu	黄文凯 HUANG Wenkai	石雪松 SHI Xuesong	张涵筱 ZHANG Hanxiao
陈铭行 CHEN Mingxing	雷 畅 LEI Chang	司昌尧 SI Changyao	周子琳 ZHOU Zilin
陈应楠 CHEN Yingnan	李宏健 LI Hongjian	王 路 WANG Lu	朱凌云 ZHU Lingyun
陈予婧 CHEN Yujing	李舟涵 LI Zhouhan	吴林天池 WU Lintianchi	
封 翘 FENG Qiao	马彩霞 MA Caixia	吴敏婷 WU Minting	
龚之璇 GONG Zhixuan	潘 博 PAN Bo	于文爽 YU Wenshuang	

2017级学生
Students 2017

卞直瑞 BIAN Zhirui	樊力立 FAN Lili	刘 畅 LIU Chang	沈葛梦欣 SHEN Gemengxin	张凯莉 ZHANG Kaili
卜子睿 BU Zirui	甘静雯 GAN Jingwen	龙 沄 LONG Yun	沈晓燕 SHEN Xiaoyan	周金雨 ZHOU Jinyu
陈佳晨 CHEN Jiachen	顾天奕 GU Tianyi	陆柚余 LU Youyu	孙 萌 SUN Meng	周慕尧 ZHOU Muyao
陈露茜 CHEN Luxi	韩小如 HAN Xiaoru	马路遥 MA Luyao	孙 瀚 SUN Han	朱菁菁 ZHU Jingjing
陈雨涵 CHEN Yuhan	焦梦雅 JIAO Mengya	马子昂 MA Ziang	杨 帆 YANG Fan	朱雅芝 ZHU Yazhi
程科懿 CHENG Keyi	李心彤 LI Xintong	彭 洋 PENG Yang	杨佳锟 YANG Jiakun	达热亚·阿吾斯哈力 Dareya Awusihali
董一凡 DONG Yifan	林易谕 LIN Yiyu	尚紫鹏 SHANG Zipeng	杨乙彬 YANG Yibin	

2018级学生
Students 2018

包诗贤 BAO Shixian	林济武 LIN Jiwu	邱雨欣 QIU Yuxin	肖郁伟 XIAO Yuwei	张 同 ZHANG Tong
陈锐娇 CHEN Ruijiao	刘瑞翔 LIU Ruixiang	沈 洁 SHEN Jie	熊浩宇 XIONG Haoyu	张新雨 ZHANG Xinyu
冯德庆 FENG Deqing	刘湘菲 LIU Xiangfei	宋佳艺 SONG Jiayi	徐 颖 XU Ying	周宇阳 ZHOU Yuyang
顾嵘健 GU Rongjian	陆麒竹 LU Qizhu	孙穆群 SUN Muqun	薛云龙 XUE Yunlong	阿尔申·巴特尔江 Aershen Bateerjiang
顾祥姝 GU Xiangshu	罗宇豪 LUO Yuhao	田 靖 TIAN Jing	杨 朵 YANG Duo	
何 旭 HE Xu	倪梦琪 NI Mengqi	田舒琳 TIAN Shulin	喻姝凡 YU Shufan	
李逸凡 LI Yifan	牛乐乐 NIU Lele	吴高鑫 WU Gaoxin	张百慧 ZHANG Baihui	

研究生 Postgraduate

艾心 AI Xin　崔傲寒 CUI Aohan　蒋佳瑶 JIANG Jiayao　李文凯 LI Wenkai　邵思宇 SHAO Siyu　王却奁 WANG Quelian　谢忠雄 XIE Zhongxiong　张本纪 ZHANG Benji　赵婧靓 ZHAO Jingliang
曹阳 CAO Yang　方飞 FANG Fei　蒋建昕 JIANG Jianxin　刘垄鑫 LIU Longxin　沈佳磊 SHEN Jialei　王晓茜 WANG Xiaoqian　徐一品 XU Yipin　张豪杰 ZHANG Haojie　周剑晖 ZHOU Jianhui
陈嘉铮 CHEN Jiazheng　冯琪 FENG Qi　蒋靖才 JIANG Jingcai　柳纬宇 LIU Weiyu　沈珊珊 SHEN Shanshan　王峥涛 WANG Zhengtao　徐亦杨 XU Yiyang　张洪光 ZHANG Hongguang　赵伟 ZHAO Wei
陈立华 CHEN Lihua　顾聿笙 GU Yusheng　姜澜 JIANG Lan　刘晓君 LIU Xiaojun　施成 SHI Cheng　王子璠 WANG Zishan　杨益晖 YANG Yihui　张靖 ZHANG Jing　周明辉 ZHOU Minghui
陈祺 CHEN Qi　胡珊 HU Shan　蒋造时 JIANG Zaoshi　刘泽超 LIU Zechao　宋春亚 SONG Chunya　吴结松 WU Jiesong　杨肇伦 YANG Zhaolun　张黎萌 ZHANG Limeng　周贤春 ZHOU Xianchun
程思远 CHENG Siyuan　黄凯峰 HUANG Kaifeng　江振彦 JIANG Zhenyan　吕秉田 LV Bingtian　宋富敏 SONG Fumin　吴松霖 WU Songlin　于慧颖 YU Huiying　张欣 ZHANG Xin　周洋 ZHOU Yang
迟海韵 CHI Haiyun　黄丽 HUANG Li　李若尧 LI Ruoyao　缪姣姣 MIAO Jiaojiao　拓湛 TUO Zhan　席弘 XI Hong　余星凯 YU Xingkai　张学 ZHANG Xue　邹晓蕾 ZOU Xiaolei
种桂梅 CHONG Guimin　贾福龙 JIA Fulong　黎乐源 LI Leyuan　彭丹丹 PENG Dandan　王敏姣 WANG Minjiao　谢星宇 XIE Xingyu

陈硕 CHEN Shuo　董晶晶 DONG Jingjing　黄子恩 HUANG Zien　梁庆华 LIANG Qinghua　聂柏慧 NIE Baihui　王浩哲 WANG Haozhe　文涵 WEN Han　徐新杉 XU Xinshan　臧倩 ZANG Qian
曹永青 CAO Yongqing　高祥震 GAO Xiangzhen　季惠敏 JI Huimin　刘刚 LIU Gang　戚迹 QI Ji　王丽丽 WANG Lili　吴帆 WU Fan　徐雅甜 XU Yatian　张馨元 ZHANG Xinyuan
陈思涵 CHEN Sihan　葛嘉许 GE Jiaxu　蒋玉若 JIANG Yuruo　刘江全 LIU Jiangquan　裘嘉珺 QIU Jiajun　王姝宁 WANG Shuning　吴家禾 WU Jiahe　杨瑞东 YANG Ruidong　章程 ZHANG Cheng
陈欣冉 CHEN Xinran　耿蒙蒙 GENG Mengmeng　金璐璐 JIN Lulu　刘宣 LIU Xuan　芮丽燕 RUI Liyan　王坦 WANG Tan　吴桐 WU Tong　杨喆 YANG Zhe　赵霏霏 ZHAO Feifei
陈妍 CHEN Yan　宫传佳 GONG Chuanjia　李鹏程 LI Pengcheng　刘姿佑 LIU Ziyou　苏彤 SU Tong　王婷婷 WANG Tingting　吴峥嵘 WU Zhengrong　于明霞 YU Mingxia　赵媛倩 ZHAO Yuanqian
程睿 CHENG Rui　桂喻 GUI Yu　李恬楚 LI Tianchu　娄弯弯 LOU wanwan　孙鸿鹏 SUN Hongpeng　王一侬 WANG Yinong　谢灵晋 XIE Linnjin　于昕 YU Xin　朱鼎祥 ZHU Dingxiang
从彬 CONG Bin　郭嫦嫦 GUO Changchang　厉伟 LI Wei　马亚菲 MA Yafei　汤晋 TANG Jin　王永 WANG Yong　熊攀 XIONG Pan　袁一 YUAN Yi　朱凌峥 ZHU Lingzheng
代晓荣 DAI Xiaorong　黄陈瑶 HUANG Chenyao　李潇乐 LI Xiaole　梅凯强 MEI Kaiqiang　童月清 TONG Yueqing　王照宇 WANG Zhaoyu　徐沙 XU Sha　袁子燕 YUAN Ziyan

曹舒琪 CAO Shuqi　贺唯嘉 HE Weijia　孙磊 SUN Lei　张培书 ZHANG Peishu　陈雪涛 CHEN Xuetao　胡慧慧 HU Huihui　刘信子 LIU Xinzi　邱嘉玥 QIU Jiayue　王瑜 WANG Yu　杨淑婷 YANG Shuting
陈辰 CHEN Chen　黄煜 HUANG Yu　王佳倩 WANG Jiaqian　张钤 ZHANG Ling　陈仲卿 CHEN Zhongqing　黄追日 HUANG Zhuiri　刘洋宇 LIU Yangyu　孙雨泉 SUN Yuquan　夏凡琦 XIA Fanqi　杨颖萍 YANG Yingping
付伟佳 FU Weijia　李子璇 LI Zixuan　王智伟 WANG Zhiwei　张彤 ZHANG Tong　程惊宇 CHEN Jingyu　孔颖 KONG Ying　刘怡然 LIU Yiran　谭明 TAN Ming　徐亭亭 XU Tingting　杨泽宇 YANG Zeyu
顾方荣 GU Fangrong　刘晨 LIU Chen　谢军 XIE Jun　张园 ZHANG Yuan　董素云 DONG Suhong　李江涛 LI Jiangtao　陆恒 LU Heng　万璐依 WAN Luyi　徐雅静 XU Yajing　张春婷 ZHANG Chunting
郭超 GUO Chao　吕童 Lv Tong　徐瑜灵 XU Yuling　章太雷 ZHANG Tailei　方柱 FANG Zhu　李汶淇 LI Wenqi　罗羽 LUO Yu　王坤勇 WANG Kunyong　薛鑫 XUE Xin　张明 ZHANG Ming
郭硕 GUO Shuo　马耀 Mao Yao　杨华武 YANG Huawu　赵惠惠 ZHAO Huihui　顾妍文 GU Yanwen　李雨婧 LI Yujing　麋泽宇 MI Zeyu　王秋锐 WANG Qiurui　杨丹 YANG Dan　赵中石 ZHAO Zhongshi
韩旭 HAN Xu　秦勤 QIN Qin　张翱然 ZHANG Aoran　周钰 ZHOU Yu　郭金未 GUO Jinwei　刘稷祺 LIU Jiqi　宁汇琳 NING Huilin　王熙 WANG Xi　杨蕾 YANG Lei
何劲雁 HE Jinyan　宋宇瑂 SONG Yuxun　张晗 ZHANG Han　陈安迪 CHEN Andi　何志鹏 HE Zhipeng　刘晓倩 LIU Xiaoqian　潘璐梦 PAN Lumeng　王晓坤 WANG Xiaokun　杨青云 YANG Qingyun

曹焱 CAO Yan　陈紫葳 CHEN Ziwei　黄健佳 HUANG Jianjia　李让 LI Rang　林瑜洋 LIN Yuyang　刘洋 LIU Yang　时远 SHI Yuan　王维依 WANG Weiyi　杨璐 YANG Lu　张文轩 ZHANG Wenxuan
潮书铺 CHAO Shuyong　迟铄雯 CHI Shuowen　黄瑞安 HUANG Ruian　李天 LI Tian　刘靖雯 LIU Jingwen　刘颖琦 LIU Yinqi　孙晓雨 SUN Xiaoyu　王熙昀 WANG Xiyun　杨云睿 YANG Yunrui　张雅翔 ZHANG Yaxiang
陈红云 CHEN Hongyun　杜孟泽杉 DUMENG Zeshan　蒋健 JIANG Jian　李晓楠 LI Xiaonan　刘恺丽 LIU Kaili　罗文馨 LUO Wenxin　孙媛媛 SUN Yuanyuan　王云珂 WANG Yunke　尹子晗 YIN Zihan　张钰 ZHANG Yu
陈健楠 CHEN Jiannan　范勇 FAN Yong　金沛沛 JIN Peipei　李星儿 LI Xing'er　刘宛莹 LIU Wanying　倪铮 NI Zheng　唐萌 TANG Meng　吴慧敏 WU Huimin　张路薇 ZHANG Luwei　张悦 ZHANG Yue
陈鹏远 CHEN Pengyuan　方园园 FANG Yuanyuan　黎飞鸣 LI Feiming　李雅 Li Ya　刘伟 LIU Wei　聂书琪 NIE Shuqi　王春燕 WANG Chunyan　夏昕宇 XIA Xinyu　张珊珊 ZHANG Shanshan　郑航 ZHENG Hang
陈启宁 CHEN Qining　冯时雨 FENG Shiyu　李谷羽 LI Guyu　梁晓蕊 LIANG Xiaorui　刘霄 LIU Xiao　秦岭 QIN Ling　王慧文 WANG Huiwen　徐琳茜 XU Linxi　张思琪 ZHANG Siqi　周珏伦 ZHOU Juelun
陈晓 CHEN Xiao　郭鑫 GUO Xin　李家祥 LI Jiaxiang　林晨晨 LIN Chenchen　刘晓芬 LIU Xiaofen　邵夏梦 SHAO Xiameng　王少君 WANG Shaojun　徐志超 XU Zhichao　张涛 HANG Tao　周郅 ZHOU Zhi
陈妍霓 CHEN Yanni　胡名卉 HU Minghui　李澜珺 LI Lanjun　林宇 LIN Yu　柳妍 LIU Yan　施少鋆 SHI Shaojun　王顺 WANG Shun　严华东 YAN Huadong　张文欣 ZHANG Wenxin　左斌 ZUO Bin

图书在版编目（CIP）数据

南京大学建筑与城市规划学院建筑学教学年鉴. 2018—
2019 / 王丹丹编. -- 南京：东南大学出版社，2020.4
　ISBN 978-7-5641-8861-0

Ⅰ. ①南… Ⅱ. ①王… Ⅲ. ①南京大学—建筑学—教
学研究—南京—2018—2019—年鉴 Ⅳ. ①TU-42

中国版本图书馆CIP数据核字（2020）第043004号

南京大学建筑与城市规划学院建筑学教学年鉴　2018—2019

Nanjing Daxue Jianzhu Yu Chengshi Guihua Xueyuan Jianzhuxue Jiaoxue Nianjian 2018—2019

编 委 会：丁沃沃　赵　辰　吉国华　周　凌　王丹丹
装帧设计：王丹丹　丁沃沃
版面制作：李宏健　陈予婧　李舟涵　周子琳
参与制作：颜骁程　陶敏悦
责任编辑：姜　来　魏晓平

出版发行：东南大学出版社
社　　址：南京市四牌楼2号
出 版 人：江建中
网　　址：http://www.seupress.com
邮　　箱：press@seupress.com
邮　　编：210096
经　　销：全国各地新华书店
印　　刷：南京新世纪联盟印务有限公司
开　　本：889mm×1194mm　1/20
印　　张：10
字　　数：475千
版　　次：2020年4月第1版
印　　次：2020年4月第1次印刷
书　　号：ISBN 978-7-5641-8861-0
定　　价：68.00元

本社图书若有印装质量问题，请直接与营销部联系。电话：025-83791830